D1156145

Rock —

I am thankful for having gotten to know you at Buckner all those years ago, and I wish you all the best for continued happiness & success.

Michael Jaye

7 July 2017

The Worldwide Flood

Uncovering and Correcting the Most Profound Error in the History of Science

Michael Jaye, Ph.D.

Copyright © 2017 Michael Jaye, Ph.D.

All rights reserved. No part of this book may be used or reproduced by any means, graphic, electronic, or mechanical, including photocopying, recording, taping or by any information storage retrieval system without the written permission of the author except in the case of brief quotations embodied in critical articles and reviews.

Archway Publishing books may be ordered through booksellers or by contacting:

Archway Publishing
1663 Liberty Drive
Bloomington, IN 47403
www.archwaypublishing.com
1 (888) 242-5904

Because of the dynamic nature of the Internet, any web addresses or links contained in this book may have changed since publication and may no longer be valid. The views expressed in this work are solely those of the author and do not necessarily reflect the views of the publisher, and the publisher hereby disclaims any responsibility for them.

ISBN: 978-1-4808-4431-5 (sc)
ISBN: 978-1-4808-4433-9 (hc)
ISBN: 978-1-4808-4432-2 (e)

Library of Congress Control Number: 2017903986

Print information available on the last page.

Archway Publishing rev. date: 05/30/2017

I dedicate this book to my wife, Donna, and to our children Megan, Matthew, David, and Emily.

Acknowledgments

None of this would have been possible were it not for the United States Military Academy at West Point, NY. Its mathematics, science, and engineering courses formed a broad foundation that I use as I try to understand the world around me; it took a while to reconcile Monterey Canyon and other geologic observations, but I have it now. Of even greater benefit, it was there that I met my wife (we are 1981 classmates), and it is where we raised our family while I was on its faculty. In addition, it was where I met Frank Giordano, a gentleman, scholar, friend, colleague, and mentor. Frank was instrumental to establishing the conditions and circumstances that would lead to this book.

I am grateful for the critique and advice from another Giordano-class individual, Jack Jensen, who helped to steel my resolve.

I am thankful for my skillful editor, Nancy Sutherland. She transformed my drafts into a more readable and accessible work. All grammatical, spelling, and style errors are mine.

I thank Kristen Tsolis for creating the original three-color Arc-GIS map like the one on the cover, and I thank Erin Greb who suggested adding to it the terrain textures. I am particularly thankful for Evan Applegate who skillfully produced the Google Earth-like maps that are vital to this work. Geologists, anthropologists, and human pre-historians will use these maps as a starting point for their disciplines' necessary reformations.

Contents

Introduction

Discovery consists of looking at the same thing as everyone else and thinking something different.

Albert Szent-Gyorgyi, Nobel Laureate

An alternate subtitle for this book could have been, "The Transformative Power of a New Observational Tool" – a chapter in Marcelo Gleiser's book *The Island of Knowledge: The Limits of Science and the Search for Meaning*. The transformational tool featured in the chapter is the telescope, and Gleiser recounts the history of one of its most famous users, Galileo, who observed that Jupiter's moons orbit Jupiter and not Earth. The ensuing transformation: the end of the idea that Earth was the center of the universe (geocentrism).

What the telescope was to Galileo in the early seventeenth century, Google Earth and Google Maps (satellite view) are to us in the twenty-first century. The maps allow us to observe the topography of submerged landscapes, something no human was able to do before the publication of the maps in the past decade or so. The information found in the new maps leads to another scientific transformation: uncovering and correcting a fundamental error in the geology that has affected the most recent two centuries of science. The error, geology's "no worldwide flood ever" paradigm, influences our understanding not only of the earth's history, but also of human history and especially anthropology. It was no small mistake.

My journey into geology began soon after I moved to the Monterey, California, area in 2009, when a long-time acquaintance suggested that I might find Monterey Canyon interesting. Out of curiosity I looked at it using Google Maps, then relatively new, and that is when I observed something quite puzzling and intriguing.

I was struck by what appeared to be a system of rivers in the offshore region (see Figure 1). How, I wondered, did those river systems become submerged in more than 3 km (about 2 miles) of water? What caused such a massive amount of water to cover them, and where did the water come from? Why didn't I know about this?

So began my education and search for the truth about geology.

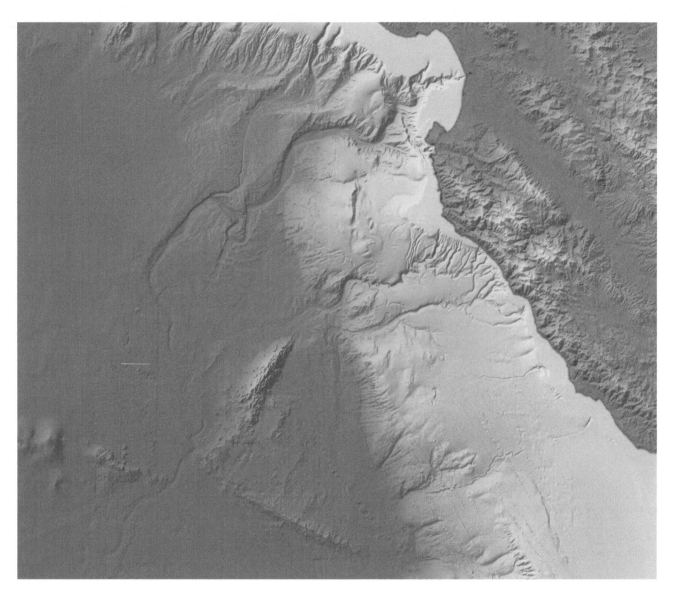

Figure 1. Monterey Canyon and the Big Sur drainages off the coast of Central California. The combined river system ends in the lower left, approximately 250 km from its origin near Moss Landing, California.

At the outset, I was unaware that because of geology, all of science dismisses the idea of a worldwide flood. I was unaware also that the story of a great flood is found not just in the Bible (the book of Genesis) but in cultures worldwide as well. Yet because of geology, all these stories are discounted as myth – at times condescendingly, but ultimately, ironically. It turns out that many of the deluge stories are preserved first-hand accounts of survivors. There was a worldwide flood, and it so transformed the Earth that it remains a part of our story.

With our inability to see ocean depths until recently, it is no wonder that geology's "no worldwide flood" tenet has lasted for as long as it has. There had been little reason to question it, despite the ubiquitous narratives found in the human oral tradition. One would think that the new bathymetry information found on the maps would have inspired geologists to discard that tenet and pursue better science (and truth). Sadly, it has not. Unlike geologists, I treat the bathymetry information found in the maps as new data, and the questions that arose when I first saw those submerged rivers inspired me to delve into a very famous, if not historic, scientific problem.

What I have found is indisputable: geology's fundamental "no worldwide flood, ever" paradigm rests on a logical error. Google Maps not only played a part in exposing that error, but they also revealed the source of a nearly incomprehensible amount of water. The maps I saw, together with other recent scientific findings, reveal that a flood inundated vast expanses of the Earth roughly 13,000 years before present (12,800 ± 150 years, with the time estimate derived from carbon dating of a nanodiamond layer caused by a cosmic impact). It nearly killed our species, and it forever transformed our planet.

Unfortunately, geologists and scientists belittle those who mention a worldwide flood. From their perspective, "no flood, ever" was such a triumph of science over religion, fact over faith, that no serious person – certainly, no scientist – would consider challenging it. The paradigm has been held so widely and for so long that it has become dogma, protected as an article of faith by geologists and scientists. I expect that adherents will attempt to dispute this book with claims that it has a religious purpose, but it does not.

In fact, this book is an exercise in science – its history and where it has brought us, and its purpose is to expose and correct geology's "no worldwide flood, ever" error. I anticipate that the findings will create a scientific revolution that, perhaps paradoxically, will restore the old as new. I use the scientific method to achieve this restoration, and I draw from such disciplines as anthropology, biology, fluid dynamics, geography, geologic processes, history, logic, mathematics, and physics. By the end of this book you will have a better understanding of who we are, where we are from, and what has happened to our planet. In addition, you are likely to agree that "no flood, ever" is the biggest error in the history of science.

I intend to make the book's materials as accessible and as understandable for as many readers as possible, because the worldwide flood is an integral and transformative part of *our* story. I hope that you will be intrigued by what follows.

Chapter 1
"No Worldwide Flood, Ever"

Bearing upon this difficult question [was there a worldwide flood?], *there is, I think, one great negative conclusion now incontestably established – that the vast masses of diluvial gravel, scattered almost over the surface of the earth, do not belong to one violent and transitory period.*

Adam Sedgwick (1831)

It might come as a surprise for many to learn that the whole of science accepts that there was never a worldwide flood. Perhaps this is the first time that you have heard it, and if that is so, then you are likely surprised because what you have heard is a famous and contradictory story, that of Noah and the biblical deluge. Compounding your surprise might be an additional piece of information: a flood tradition appears in cultural narratives from around the world – the story of Noah has hundreds of cousins. All are discounted as myth because of geology.

Interestingly, most modern, lettered, elite geologists are unaware of the source of their "no flood, ever" doctrine. Today's geologists know neither the argument nor the facts that led to their foundational conviction that there was never a worldwide flood. They simply accept it. Ironically, the few who know the history of the debate call it a triumph of science over religion. Having survived unchallenged for nearly 200 years, the belief has evolved into a fundamental tenet in geology; in fact, an entire subdiscipline, submarine geomorphology, rests on it. Perhaps geologists have no reason to understand the history of their "accepted science" because the finding has been sustained for such a long time; and everyone practicing science simply accepts it as well.

Why did they think there was no worldwide flood? I have found that "no flood, ever" has a history that is easily distilled. It finds its origin in the early decades of the nineteenth century when geologists in Europe debated whether the whole of the earth suffered a deluge. Geologists set about various parts of the continent and discovered that diluvial deposits (sediments carried by floodwaters) belonged to multiple distinct events. This finding indicated that all those boulders were not transported by a single flood. And, in fact, some were found to have been transported by glaciers. Thus, at its essence, the argument against the worldwide flood went like this: because there was no common event in the diluvial records, there could never have been a single worldwide flood.

Several key figures influenced the debate, among them the Reverend Adam Sedgwick, Woodwardian Professor at Cambridge University, and for two years president of the Geological Society of London. In his farewell presidential address at the society's 1831 annual meeting, Sedgwick recanted his belief in the single flood (italics added):

> The vast masses of diluvial gravel ... do not belong to one violent and transitory period. It was indeed a most unwarranted conclusion when we assumed the contemporaneity of all the superficial gravel on the earth. ... *Having been myself a believer* [in a worldwide flood], *and, to the best of my power, a propagator of what I now regard as a philosophic heresy, ... I think it right ... thus publicly to read my recantation.*

This and other remarks from a portion of the farewell address are provided in Appendix A.

Sedgwick's standing as the Society's president, as a Cambridge University professor, and as a reverend played an important role in imparting lasting effect to his recantation. Moreover, several of Sedgwick's distinguished contemporaries came to a similar conclusion, and their collective prestige ensured permanence. Geology's "no flood, ever" paradigm persists to the present.

It is an understatement to say that "no flood, ever" has had a profound effect not only in geology but also in associated disciplines such as anthropology, archaeology, and human history. "No flood" was a historic, celebrated, and influential finding.

An example of its exalted status is found in Stephen Gould's *The Flamingo's Smile, Reflections in Natural History* (1985), in which he praises Sedgwick's scientific spirit (italics added):

> He [Sedgwick] had led the fight for flood theory; but he knew by then [1831] that he had been wrong. He also recognized that he had argued poorly at a critical point: he had correlated the caves and gravels not by empirical evidence, but by a prior scriptural belief in the Flood's reality. As empirical evidence disproved his theory, he realized this logical weakness and submitted himself to rigorous self-criticism. *I know no finer statement in all the annals of science than Sedgwick's forthright recantation ... it illustrated so well the difference between dogmatism, which cannot change, and true science. ...*

Chapter 2
The Historic Error

But first generalizations are almost always pushed too far. After being bewildered with the observation of unconnected facts, the first glimmering of general truth is so delightful, that it often leads us beyond the bounds of fair induction.

Adam Sedgwick (1831)

In drawing their "no worldwide flood, ever" conclusion, Sedgwick and his contemporaries never considered that landscapes now submerged might once have been exposed. Nor were they able to look into the depths to discern their topography and morphology (the configuration and evolution of landforms). Instead, they assumed that evidence of the flood would be found in presently exposed landscapes, which is not necessarily true, for the flood might have filled the abyss from the depths upward (which is what actually happened, as we will see). These early geologists assumed that the present amount of water has been with the earth since its beginning, thereby precluding the possibility that now-submerged landscapes were at one time exposed but were later inundated as a result of some unknown event.

The early geologists' precise conclusion from the evidence should have been that *a worldwide flood did not inundate presently exposed landscapes.* This is completely and undeniably true, and no one can or would argue against the fact that where we are now (i.e., on the presently exposed landscapes) was never submerged by a common flood event. That a worldwide flood did not inundate presently exposed landscapes is precise and correct, yet it is wholly different from geology's claim that there was never a worldwide flood. Sedgwick and his peers went too far. They exceeded the bounds of fair induction.

Let us be clear: because presently exposed landscapes were never subjected to a worldwide flood does not in any way imply that now-submerged landscapes have always been that way. Unfortunately for science, and especially for our understanding of human history, the early geologists' incorrect conclusion became an accepted, fundamental belief.

It is indisputable that "no worldwide flood, ever" is an incorrect conclusion. Alone, recognition of this historic error should cause any geologist, scientist, or historian to reflect and consider its potentially profound implications, for it is possible that a significant portion of all subsequent findings are adversely affected. Two hundred years of science is potentially in error.

Being a scientist myself, I brought this matter to the attention of geologists associated with the discipline's most prestigious journals, and what I discovered is that to geologists, "no flood, ever" is an irrefutable, accepted fact. That otherwise intelligent individuals fail to recognize that their science rests on an error demonstrates the constraining power that "no flood" holds over science. It is dogma, adhered to and protected as an article of faith, with its highest-ranking practitioners not knowing (or caring to know) the history of their core belief. This is as astonishing as it is disappointing.

At this point I do not want to commit the Sedgwick-like error of "going too far." Merely exposing geology's foundational error does not prove that there was a worldwide flood, although it does afford the possibility that there was one. Likewise, exposing the error bolsters the validity of the many cultural narratives about a devastating flood. More important, however, is that the logical error and its potentially far-reaching consequences immediately diminish geology's standing as an authoritative science.

I have encountered geologists' blind adherence to their faith. And so Stephen Gould's words of praise cause me to wonder where the modern Sedgwick is? Does there exist a geologist with similar courage? Who among geologists will recognize the logical error upon which their erroneous "no flood, ever" paradigm rests? Which geologist will submit the discipline to rigorous self-criticism?

While waiting for this individual to come forward, we can ponder the following: had the early geologists not erred, had they not celebrated the incorrect finding, and had they waited until they could observe the topography of the depths, then we might not have suffered the 200-year dichotomy between science and the human narrative tradition.

Chapter 3
Submerged River Systems, Geology's Error, and the Scientific Method

The fact that an opinion has been widely held is no evidence whatsoever that it is not utterly absurd; indeed, in view of the silliness of the majority of mankind, a widespread belief is more likely to be foolish than sensible.

Bertrand Russell (1929)

Post-Sedgwick geology exemplifies the structure of scientific revolutions (described by Thomas Kuhn in his famous book by that title). Applying Kuhn's tenets to geology, we realize that "no worldwide flood, ever" began with a collection of facts that showed there was no common flood event (in presently exposed landscapes). This discovery created a crisis, as Sedgwick explained in his farewell address, and it would end with the erroneous theory overturning the historical narrative(s) about a worldwide flood. The evidence and conclusions drawn in support of "no flood, ever" were published in nascent scholarly journals of various geological societies, gaining acceptance as the basis for other geologic findings, and eventually attaining paradigmatic status in academia and elsewhere. Normal science ensued, and "no flood, ever" persisted; scientists use it as a basis for knowing what our world is like. Nearly 200 years of published research has been fit into this "scientific" worldview.

All of this is quite unfortunate. We now recognize that geology's foundational tenet went too far and instead should have allowed for the possibility that a flood once inundated some part of now-submerged landscapes. The big question is why geologists have not recognized the error, not only because of its historical import but also because they have the same (or better) access as the rest of us to maps that reveal the formerly unknown, submerged landscapes.

Several examples of submerged river drainage systems are shown on Figure 2. Viewed at their proper level, the submerged structures found on the map images prompt the question, *How could these features all exist in more than 3 km of water if there was never a worldwide flood?* I will take some time addressing this question.

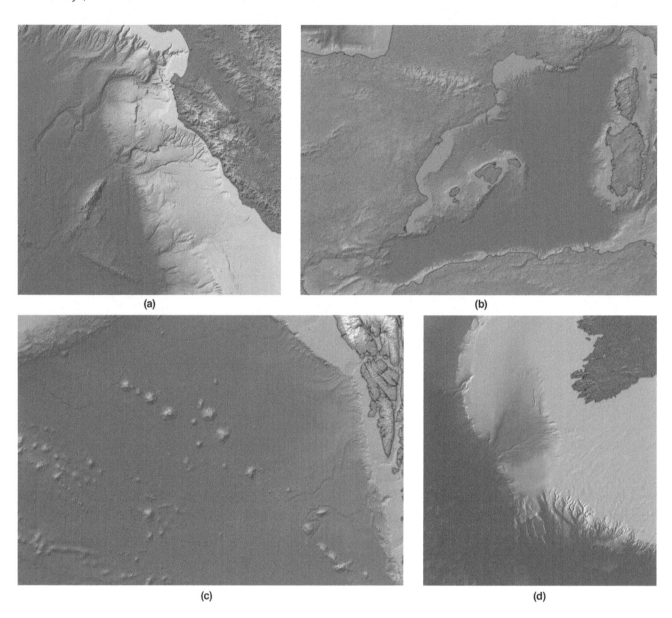

Figure 2. Former river systems once flowed from upland regions, through shelves, into abyssal plains, and ultimately to large basins and reservoirs. Examples include: (a) Monterey Canyon and the Big Sur drainages off the coast of California. The combined system ends in the lower left of the image, approximately 250 km from its origin near Moss Landing, California. (b) The western Mediterranean Sea and the many river drainages that flowed into a large central reservoir. The length of the prominent, westernmost submerged river drainage in this image is nearly identical to that on Figure 2a. Note that the drainages shown on Figure 2b share a common terminal depth, approximately 2,600 m below present sea level. (c) Several submerged drainages in the Gulf of Alaska. The origin of the northernmost system can be identified in presently flowing systems in Canada; the river flowed westward down the continental shelf and then nearly 700 km through the abyss toward what is now the Aleutian Atoll. Another system in the southeast of the image meandered between volcanoes. (d) The now-submerged system in the Celtic Sea, found in the center of the image, originated approximately 400 m below present sea level and ended in a location nearly 3,000 m deeper. The drainage system found in the lower part of the image (resembling inverted tree roots) had a much steeper gradient that created wider channels than those in the other nearby system.

First, it should come as no surprise for us to learn that the submerged structures represent inexplicable anomalies to geologists, who cannot account for their formation in the context of "no worldwide flood, ever." In a recent paper published in the *Journal of Sedimentary Research*, Metivier et al. (2005) drew attention to the problem (italics added for emphasis):

> *Yet the existence of submarine canyons and meandering channels in the deep sea remains essentially beyond our understanding.* Are these channels produced by huge catastrophic turbidity currents, or perhaps by hyperpycnal flows [when denser water flows beneath basin water forming a sediment-laden density current] or by steady-state currents formed at the mouths of large rivers? Were they initiated as rivers during past glaciations when sea level was lower? Can present-day currents recorded in some deep-sea channels, like the one in the Zaire fan, account for such structures?

> *The debate is not closed, mainly for two reasons: the difficulty in making measurements and observations on active or abandoned channels, and the probably long time scale, on the order of thousands of years, needed to develop these structures, that forbids the observation of processes on a human time scale.*

> Because of these drawbacks, researchers have focused on *experimental studies of gravity and turbidity currents* and numerical simulations.... *Experimental models* driven by stratigraphic studies have long concentrated on turbidity currents and the deposits they form. . . .

> *Experimental knowledge* on the dynamics of turbidity surges and currents was then used by modelers to propose mechanisms of channel incision and meandering in the submarine environment. . . . *Because of the lack of knowledge on submarine erosion* most of the experiments and models have been aimed at the reproduction of one-dimensional profiles, whether topographic or stratigraphic. . . . But a physical basis for channel incision and evolution in submarine environments still remains to be tackled because of the apparent difficulty in reproducing subaqueous analogs of submarine channels.

Of all scientists, geologists more than any others should know that what we see on the maps are submerged rivers, canyons, and drainage systems that could only have been created by well-understood subaerial processes – they were carved and eroded by moving water exposed to the atmosphere. But if geologists must conform to "no flood, ever," then explaining the structures is beyond their ability, as Metivier mentions. The formation process remains inexplicable simply because there is no physical process by which these features could have formed underwater.

Yet submarine geomorphologists try. On several occasions when scientists have observed turbidity flows (roiling, energetic, sediment-filled, gravity-induced slides) around some of the submerged structures, they have hypothesized that such flows constitute their creation mechanism. Rather than question their foundational belief as the new map data ought to have prompted them to do, geologists instead want us to believe that these rarely observed turbidity flows somehow persist over hundreds of kilometers in more than 3 km of water to create these massive structures? But if that were so, wouldn't we observe these flows more frequently? Wouldn't they be omnipresent in the main channels?

By what process could the central channels have formed, and how could river-like sediments be found in their beds? Shouldn't bed sediments have been scoured away by energetic flows merging from feed channels? How could these flows exist in abyssal plains where we observe many of these former river systems? What is the mechanism for their formation and the energy source for their propagation? By what physical process could these fast-moving energetic flows merge (exactly as rivers)? And how could meanders have formed in the presence of flows that fall straight down gravitational gradients?

These are all rhetorical questions, for turbidity flows as the mechanism to create these structures are but ad hoc explanations conjured to fit the prevailing paradigm. Whatever turbidity flows have been observed in the vicinity of these submerged geologic formations are simply post-submersion consequences and certainly not their creation mechanism. Geologists have conflated cause and effect. In the context of the scientific method, what they are attempting to do is fit the data (what we observe on the maps) to their "no flood, ever" theory. Fitting data to theory is not science, but rather it is anti-science, or fantasy. While turbidity flows do certainly exist, ascribing them as the mechanism by which these submerged structures are carved is pseudo-science.

In the presence of new information that contradicts accepted theory (already shown to be in error), the scientific method would have us reject the theory. Geology has failed to do this because the "no worldwide flood, ever" dogma has hindered its believers from fulfilling their responsibility to be skeptical, objective scientists. The new map data should have caused geologists to pursue the truth, discover the error, and then reject their erroneous belief.

Some defenders of the geologic faith might try to argue that the longevity of the "no flood, ever" theory testifies to its correctness, but that would be to grasp at straws. The erroneous theory has persisted only because we could not see into the depths until quite recently and because it is undeniably true that where we are now was never flooded by a single, common event. In essence, "no flood, ever" lasted for so long because of confirmation bias and lack of contradictory evidence.

Again, we can ponder: had the early geologists not erred, had they afforded the possibility that what is now submerged was at one time exposed, and had they waited until we could see into the depths, then the discovery of the submerged drainage systems would have led to immediate corroboration of the human narrative tradition. Their conclusion would have been obvious, namely, that there was a worldwide flood.

Chapter 4
Contradictions and Inconsistencies Due to the Historic Error

Reasoning must never find itself contradicting definite facts; but reasoning must allow us to distinguish, among facts that have been reported, those that we can fully believe, those that are questionable, and those that are false.

Rene Reaumur, 18th century French scientist

I have uncovered the poor reasoning that led to geology's erroneous "no worldwide flood, ever" tenet, and I have discussed its ensuing pseudo-science crafted to fit new observations (the submerged structures found all over the world) into that error. Yet among the consequences of pseudo-science is that contradictions and inconsistencies eventually emerge. In geology they are found in published works dealing with the formation mechanism of the submerged topographic features. One inconsistency is attributing the morphology of submerged features in the western Mediterranean Sea to subaerial process while simultaneously attributing the morphology of all other similar features (like those on Figures 2a, 2c, and 2d) to turbidity flows. One contradiction is found regarding alleged subsurface erosive behaviors: carving by turbidity flows did not exist in the western Mediterranean Sea, yet everywhere else these flows allegedly carved the myriad features we now see in the bathymetry. Another inconsistency is found in the shape and dimensions of nearly identical canyons, one subaerial and one submerged nearby.

The new maps show that an inland sea existed in the western Mediterranean Sea before the flooding that brought it to its current depths. We can infer this former sea's existence from the map images. For instance, on Figure 3a we notice that submerged canyons and drainage systems ring the perimeter of the western Mediterranean Sea and that all of them terminate at a common depth. That common depth, roughly 2,600 m below present sea level, allows us to discern the former shoreline, which is approximated by the black outline on Figure 3b.

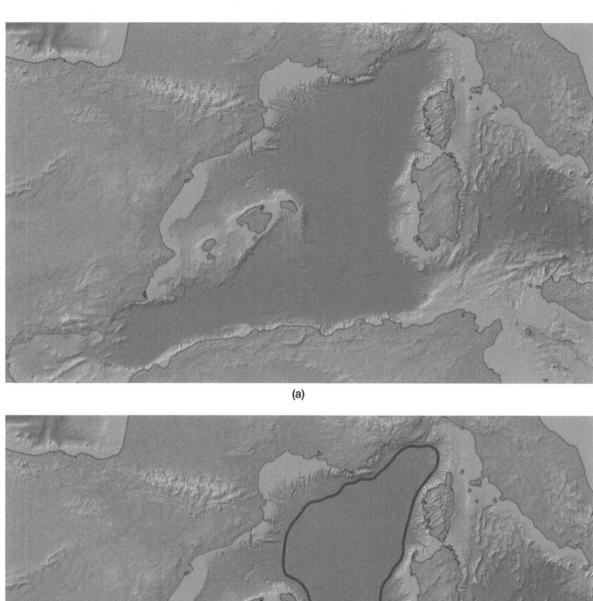

(a)

(b)

Figure 3. (a) The western Mediterranean Sea and the many river drainages that once flowed down surrounding shelves ringing the basin and into a large central reservoir. Note that all the drainages share a common terminal depth, approximately 2,600 m below present sea level. (b) The black curve approximates the former shoreline.

Interestingly, these canyons and drainages are accepted by geologists as having been subaerially carved and then preserved in the bathymetry when the Mediterranean Sea flooded through what is now the Strait of Gibraltar. A paper by Garcia-Castellanos et al. (2009) published in the journal *Nature* signifies geology's affirmation that (1) a reservoir once existed in the basin's depths; (2) the sea must have been exposed for a period of time on the order of tens of millions of years, during which flowing waters carved the surrounding canyons and riverbeds; and (3) all the structures became preserved in the bathymetry when the Mediterranean Sea flooded through the Strait of Gibraltar. (It should be noted that Garcia-Castellanos et al.'s timing of the flood event is grossly incorrect.)

The Garcia-Castellanos et al. paper has several important implications. First, it is an admission that the now-submerged drainages were subaerially carved as they sought their terminal basin. Second, once the drainages' waters met their terminal basin, their ability to erode ceased despite substantial depths through which erosion might otherwise have continued. Each is well understood: waters flow and erode subaerially, and, upon meeting their terminal basin, waters' erosive capacity halts because of a lack of gravitational potential and a change in momentum.

The inconsistency: whereas geologists accept that the Mediterranean Sea flooded through the Strait of Gibraltar, thereby preserving its subaerially carved drainages, they cannot apply a similar explanation to the formation of any other systems, because to them there was never a worldwide flood. Do geologists believe that all non-Mediterranean systems were carved by phantom subsurface processes? That would be contradictory because if turbidity flows carved all these other formations, then why did the erosive process cease at a common depth in the Mediterranean Sea when sufficient depths existed through which subsurface erosion could have continued?

The obvious answer is that like those systems ringing the western Mediterranean Sea, all the other river drainage systems and canyons found on the planet were subaerially carved while their waters sought terminal basins. We find them well preserved in the bathymetry because they were quickly submerged by the worldwide floodwaters. It is that simple.

Another inconsistency is found off the coast of Central California near the Big Sur region, a part of which is shown below on Figure 4. Two nearly identical valleys whose heads are only 4 km apart are shown circled, one far below the present ocean surface and the other subaerial. It is important for us to note that under geology's prevailing and erroneous belief, the submerged valley is far too deep to have been uncovered during any previous ice age. In addition, the maps allow us to discern that the valleys have nearly identical dimensions (vertically and horizontally). Under the "no flood, ever" doctrine, the dimensional similarity would imply that subsurface erosive processes are identical to subaerial erosive processes, an absurdity when one considers the immense density and viscosity differences between the media.

Figure 4. Two nearly identical valleys (circled) in the Big Sur region of Central California are only 4 km apart, but one is onshore and the other is submerged. The valleys' slopes, derived by measuring the change in elevation of the valleys' central axes and dividing it by the nearly north-south central axes' horizontal distances, are essentially equal.

The only way that the two features could be so similar is for both to have been subaerially carved over eons of essentially identical weather exposure, but only the lower western canyon became submerged by the worldwide flood's waters.

Our charge now is to identify the source of so much water that submerged all the subaerially carved canyons and river drainage systems existing in the earth's bathymetry.

Chapter 5
The Worldwide Flood

For of man, and the works of his hands, we have not yet found a single trace among the remnants of a former world entombed by these ancient deposits.

Adam Sedgwick (1831)

As we begin our search for the source of the floodwaters, we must recognize that the volume of water necessary to cover the former river drainage systems and canyons by more than 3 km cannot be stored in frozen form on earth. There is not enough room at the poles; nor is the atmosphere thick enough to afford the deposition of that much snow and ice. The atmosphere extends only so far, and the amount of water necessary to cover the submerged systems is so voluminous that the source must be from somewhere other than the earth – it must be from a cosmic impact. To some the idea of a cosmic impact recent enough to be within human memory might sound far-fetched, but the evidence is at hand. New maps will reveal the impact site, and its timing and effects will be corroborated by several recently published papers, as well as by other observations and scientific works.

I should also address the nature of comets, which brings us to NASA's Deep Impact mission that probed comet Tempel 1 to determine its composition. The mission's findings provide very important information that will have immediate application as we analyze impact effects and the volume of water delivered. Also vitally important to what lies ahead is that, according to Deep Impact's principal investigator, Dr. Michael A'Hearn, comets are porous objects, mostly open space, "unbelievably fragile," and "less strong than a snowbank" (Wilson 2005).

Though the object that struck the earth was not a comet, those comets that we observe are but tiny fragments of the type of object that did. This should not surprise us since small comets like Tempel 1 are asymmetric, jagged, and non-spherical, characteristics that imply fragmentation. Furthermore, we expect a spherical parent object for these fragments due to amalgamating gravitational forces from some central, attracting nucleus; case in point: Deep Impact found no discernible, solid central nucleus in Tempel 1.

Another necessary piece of background information involves what geologists recognize as the Younger Dryas (YD) event. It is an episode marked by abrupt increases in snowfall and dramatic changes to flora, fauna, climate, and the oceans (Firestone et al. 2007). Its precise cause is unknown, although it has been attributed by some to a cosmic impact that has yet to be identified – that is, until this chapter.

To the matter at hand: two recent papers from geology conclude that a cosmic impact roughly 12,800 years before present not only caused the YD effects (Kennett et al. 2015), but it also formed an associated layer of nanodiamonds (microscopic diamond crystals that are created by very high-velocity collisions) found across most of the planet (Kinzie et al. 2014). Interestingly, none of the papers identifies the impact, something that we accomplish immediately.

The impact remnants are found in the Southern Ocean southeast of South Africa, north of Antarctica, and south of Madagascar; the impact center is in the vicinity of 57°S, 53°E. Figure 5 shows three views of the impact site from a common, fixed perspective roughly 8,000 km above sea level: the standard map image (top), a bathymetry overlay (middle), and a magnetic anomaly overlay (bottom, Korhonen et al. 2007). The mostly submerged remnant crater measures approximately 2,500 km in diameter, indicated by the super-imposed line on Figure 5 (top).

Entry effects that broke off portions of the fragile impacting object (IO) account for the gap in the center of the crater's crescent. Impact craters in North America (e.g. Carolina Bays) and South America were created by IO ice fragments that rained down along the IO's path just prior to impact. (A discussion of the ice craters is found in Appendix B.) Some minerals with positive magnetic susceptibility introduced by the IO were projected nearly 1,500 km to the north and northeast through the crescent gap by impact velocities and associated forces (Figure 5, bottom). What appear to be parallel central scrapes emanating from the impact center (Figures 5, top and middle) are actually the sides of a trough measuring 1,000 km in length that was carved by the dense nucleus as it skidded northward. This trough corresponds to a band of intense magnetic anomalies (red stripe on Figure 5, bottom) created from materials worn from the nucleus during its immediate, post-impact transit. At the end of the trough are the IO's nucleus materials (circled region on Figure 5, middle) that served as the gravitational sink needed to attract and aggregate the outer ice and debris layers in the Oort Cloud where the IO is likely to have formed. Raised regions interior to the crescent are deposit mounds (yellow, orange, and red regions on Figure 5, middle), remnants from the melted mineral-ice complex that comprised the IO's outer layer. These mounds also correspond to regions of intense magnetic anomalies (Figure 5, bottom).

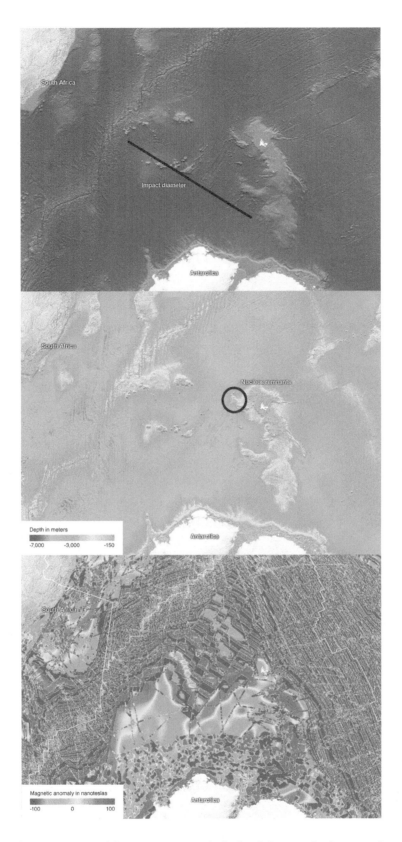

Figure 5. Identical perspectives of the impact site include: (top) the standard view with a superimposed diameter measuring 2,500 km; (middle) a bathymetry map depicting raised regions of IO-borne deposits with a superimposed circle identifying nucleus remnants; and (bottom) overlay of magnetic anomalies, from Korhonen et al. (2007), that were created by the impact and its deposited minerals. Magnetic anomalies extend approximately 1,500 km to the northeast through the "crescent" gap.

Our understanding of the origin and evolution of the universe could be informed by the age and composition of minerals obtained from the impact area's deposit mounds. In addition, cosmologists, astronomers, and other scientists might be interested in the composition of the IO nucleus materials, especially since Earth's waters are known to be older than the solar system, which might mean that the solid nucleus could be older than the solar system, too. With that in mind I submitted proposals to the International Ocean Discovery Program (IODP) to recover and analyze these materials. One reviewer commented, "It is thinking like this that moves science forward." Indeed. Unfortunately, however, the chair of the IODP is not only a geologist but also a submarine geomorphologist whose academic life rests on the flawed assumption that there was never a flood. My proposal was denied. Though initially disappointed, I am confident that one day the mission to recover parts of the IO nucleus will be successfully accomplished. I look forward to learning of its age and composition.

As depicted on Figure 6, the IO had a dense, solid core that was surrounded by a fragile outer layer. Having formed in the Oort Cloud, this outer layer was consistent with known comet composition, which as noted above, is porous, mostly open space, "unbelievably fragile," and "less strong than a snowbank."

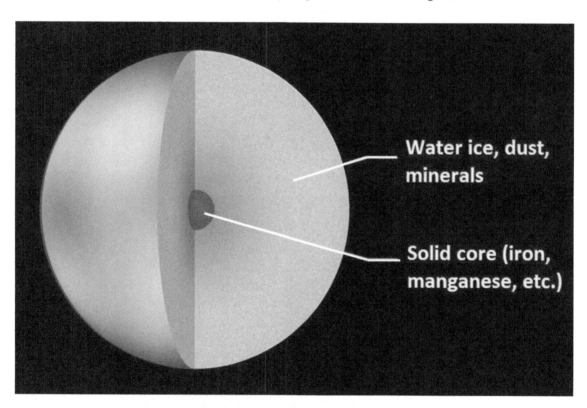

Figure 6. The impacting object had a dense core surrounded by a layer composed of water ice, rocks, and minerals.

There are immediate consequences to cosmology and astronomy. As suggested earlier, the smaller, irregularly shaped comets we observe are but fragments of these larger objects that have broken off due to gravitational interactions. Short period comets, especially Jupiter-family comets, are very likely fragments shorn from the IO during its Earth approach. In addition, Pluto could very well be a member of a class of objects similar to the IO.

It is interesting to consider the appearance of the IO as it neared Earth impact, particularly in the context that we can see small comets (Halley's comet measures roughly 15 km on its major axis) from millions of miles away. In contrast to Halley's comet, the IO was on the order of 10,000 times larger by surface area and 1,000,000 times larger by volume. Similarly composed, then the illumination from the nucleus and its tail as it approached Earth must have been frightening and memorable, particularly since the flood ensued nearly immediately after its disappearance. Therefore, it is no surprise that we find recollections of the IO in ubiquitous, ancient oral traditions. It is known by names such as Phaeton, Typhon, Set, Ta-vi, and Satan.

Pliny the Elder described Phaeton's approach: "A terrible comet was seen by the people of Ethiopia and Egypt. . . . It had a fiery appearance and was twisted like a coil, and it was very grim to behold; it was not really a star so much as what might be called a ball of fire." (Rackham, 1938) According to Allan and Delair (1995), Phaeton "was anciently regarded as a generally round, brilliantly fiery body of appreciable size, and much more star-like or sun-like than conventional comets: and it was held to have in some way caused the Deluge." The fiery comet-like appearance of the IO as it neared Earth impact and the irreversible changes induced by its flood likely account for the long-held notion that comets are harbingers of change. In addition, the Chinese New Year water dragon, a glowing, fiery serpent depicted above the clouds with water emanating from its mouth, is very likely a commemoration and memorialization of the IO's frightening appearance and effects.

In the Oort Cloud where the IO formed, gravitational accelerations induced by the solid inner core were less than 1% of Earth's accelerations. Aggregations forming the IO's outer shell of water-ice and debris made it more massive than its core, yet accelerations induced by the entire object were on the order of 2% of Earth's (see Appendix B for the acceleration derivations). These comparatively small accelerations account for the IO being so loosely packed, porous, and fragile; its gravitational accelerations were too small to act as a compressing mechanism. This corroborates A'Hearn et al's characterization of comets as less strong than a snowbank. Furthermore, the IO's fragility explains why impact effects were far less damaging than what otherwise might be expected from a similarly sized yet solid object. It was like a lightly packed but fast-moving snowball – with a rock in the middle – hitting a brick wall.

Despite its delicate outer layers, the IO, with its impact forces, nonetheless created the impact crescent that we observe in the maps, and it shifted local topography along the western edge of the crescent by up to 150 km; terrain elongations are evident in map images of the impact crescent's western extent.

It is likely that the IO was displaced by a binary star system that passed through the Oort Cloud roughly 70,000 years before present (Mamajek et al. 2015), was captured by our sun's gravitational field, and was then brought into Earth's path. Upon impact, collisions and interactions between energetic IO-borne minerals and terrestrial materials created the YD nanodiamond layer, placing the impact approximately 13,000 years before present (Kinzie et al. 2014). The timing and planet-changing consequences of the event have been preserved in the human oral tradition (the last 200 years notwithstanding).

With a diameter of 2,500 km, the IO occupied a volume of 5.58 * 10^9 km³. Given that it was composed as Tempel 1, that is, 75% open space, 2/3 of its mass pure water ice (A'Hearn et al. 2005; Kerr 2005; Sunshine et al. 2007), then 1/6 of the sphere's volume would be ice. But that ice melted, so we must account for the slight volumetric difference between ice and its melted form; thus, the IO's equivalent water volume was 1.29 * 10^9 km³. To approximate the equivalent depth of water delivered, the volume can be divided by the present oceans' surface area. Since the earth's oceans are reported to cover 3.62 * 10^8 km², *the IO delivered an average ocean water depth of 3.57 km (more than two miles).*

The newly introduced waters flooded the planet, and they did so from the former abyss upward; the floodwaters did not inundate presently exposed landscapes. (Immediately after the impact, a worldwide increase in relative humidity likely caused incredible and memorable rainfall that some survivors assumed to be the flood's source.) Coupled with the vast new waters and ensuing changes to weather patterns, the IO induced irreversible ecosystem and climatic changes that geologists recognize as the YD event. In short, the worldwide flood and the YD event are synonymous.

That the worldwide flood and the YD event are synonymous is corroborated by a recent finding in archaeoastronomy wherein an analysis of pillar carvings at Göbekli Tepe "provide evidence that the famous 'Vulture Stone' is a date stamp for 10950 BC ± 250 yrs, which corresponds closely to the proposed Younger Dryas event." (Sweatman and Tsikirsis, 2017) The study also notes that the people of Göbekli Tepe remained interested in the event several thousand years afterward, suggesting that "it had a significant impact on their cultural development." The struggle to survive and adapt to the post-flood ecosystem so affected the Göbekli Tepe culture that they etched the event's memory in stone.

Eventually, geology will recognize that the flood so dramatically transformed the earth that, by definition, it brought on a new geologic era. Let us call it the Post-Diluvian.

Indisputable proof of the worldwide flood will be attained once we discover the works of man entombed in ocean depths. I will argue that such evidence has already been found – traces of pre-flood Earth exist, some of which are found on the new maps.

For now, let us consider Earth before the flood, as well as other evidence of the flood's occurrence.

Chapter 6
Earth Before the Flood Occurred, and Other Flood Evidence

When a thing is new, people say: "It is not true." Later, when its truth becomes obvious, they say: "It is not important." Finally, when its importance cannot be denied they say: "Anyway, it is not new."

William James

We know now that a recent cosmic impact delivered the worldwide floodwaters. Therefore, let us look at something old with new eyes: Earth before the flood.

Figure 7 is an ArcGIS-produced model of pre-flood Earth with an estimated average of 3 km less water coverage than the present sea level. The figure shows that, before the flood, formerly exposed landscapes (tan and beige) were more abundant than those submerged (blue); vast seas and oceans existed before the flood, but they were disjointed. Presently exposed landscapes (beige) were more than 3 km above the former sea level, and continents existed in places now covered by our extensive oceans. The importance of this map cannot be overstated, for it allows us to notice something that is not even visible that we had never before considered.

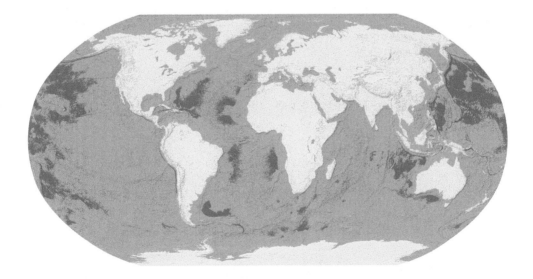

Figure 7. With more than 3 km of water graphically removed, a model of land and sea distributions in pre-flood Earth shows previously exposed but now-submerged landscapes (tan), presently exposed landscapes (beige), and former oceans and seas (blue).

The removal of so much water affords the annotation on Figure 8 of the Monterey Canyon map presented in the Introduction. What is now California would have been continuously inundated by rains induced by winds uplifted by the nearly vertical and formerly exposed continental margin. Eventually those rainwaters would be energized by the more than 3 km fall down the shelf, and their scouring interactions would eventually create Monterey Canyon. Identical processes account for the many well-preserved river drainage systems found submerged all over the planet.

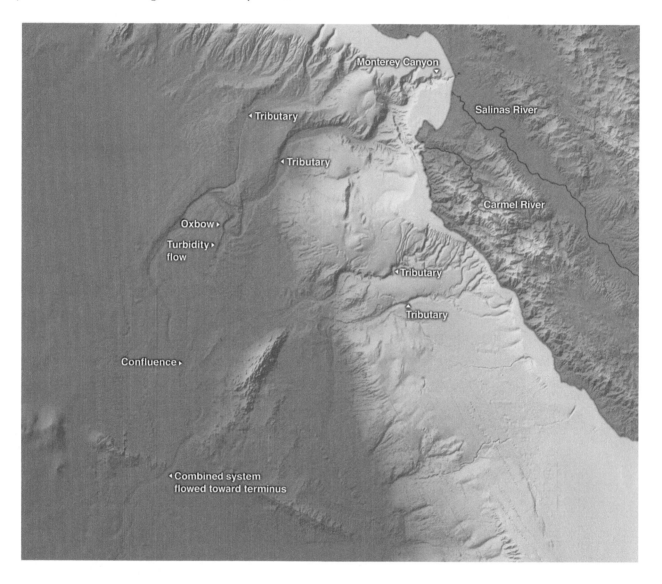

Figure 8. Annotated map of the Monterey, California, region showing features now offshore in the Pacific Ocean. The image depicts what are now subaerial intermittent streams (Salinas and Carmel Rivers), many now-submerged tributaries, an oxbow or meander, a massive turbidity flow, and the confluence region of two tributaries wherein the tributaries' beds briefly lose their definitions.

The combined Monterey Canyon and Big Sur drainages eventually flowed into a collection basin near the system's terminus approximately 250 km from what is now Moss Landing, California (located very close to Monterey Canyon's source and just north of the Salinas River's terminus on Figure 8). After its fall down the shelf, and as it neared the abyssal plain, the Salinas River system carved a prominent oxbow that is 8 km in diameter and located approximately 80 km from the present shoreline.

The straight trail left by a massive landslide caused by the collapse of the oxbow's southwestern wall is also evident. The collapse was caused by rising ocean waters impinging on the riverbank that had become weakened as the river swelled with rainfall resulting from the cosmic impact. We note that the turbidity flow fell straight down the gravitational gradient and that it did not attempt to organize itself into other pre-existing flows in the area; in addition, its remnants lack any semblance to the other submerged river-beds in the region. Finally, we note that the region to the left side of Figure 8 is in the abyssal plain where geologists' presumed gravity currents could not exist due to the absence of sufficiently steep gradients.

The confluence region depicted on Figure 8 appears somewhat ambiguous or smeared as a consequence of river-borne sediments being deposited into rising floodwaters much like the formation of river deltas. Other river-borne materials that were deposited into the rising ocean waters account for the region's sediment-filled channels (Fildani and Normark 2004).

The National Oceanic and Atmospheric Administration (NOAA) has obtained core samples from this confluence region. NOAA reports that the primary composition of the cores' materials is terrigenous sands, and the secondary composition is terrigenous gravel deposits (NOAA 2011). In other words, the sediments taken from the confluence region located more than 80 km from shore (and now in more than 3 km of water) are derived from terrestrial environments, not marine environments. Indeed they were: the sediments were carried and deposited by the pre-flood subaerially flowing river, and they were deposited when its waters met the newly introduced, rising ocean level.

The more northern of the two tributaries indicated on the right side of Figure 8 drained what is now the Big Sur region westward and then to the north of what is now a seamount. This river's course through abyssal region is somewhat difficult for us to discern on the map because, like the confluence, it is filled with sediments. Since this is a relatively flat region (there was only a 120 m elevation drop over the 40 km it traveled from the shelf toward the confluence), the riverbed is smeared or ambiguous on the map because the abyssal plane through which it flowed became filled by river-borne sediments deposited into newly rising ocean waters.

Figure 9 is a map of a portion of the Salinas Valley in Central California. It demonstrates not only the local ecosystem before the flood but it also reveals environmental changes caused by the worldwide floodwaters. The flat region in the central part on the map was once a lake bed, and the former water level can be discerned on each side of what is now the valley. The lake and its surrounding region were more than 3 km above the former sea level, and uplifted winds condensed to create persistent rains that eroded the hills and filled the lake. These waters would eventually drain to the northwest and then down the shelf into Monterey Canyon. The annotation on the map identifies the only post-flood intermittent stream in the valley. It is carving its way westward toward the Salinas River as it drains the series of arroyos and hills to the east and northeast. It demonstrates to us what 13,000 years of erosion looks like.

Figure 9. A post-flood intermittent stream has carved into the Salinas Valley from the surrounding hills and their drainage systems.

Figure 10 is a photograph of the Monterey Peninsula shoreline region. Rocks in the surf, those protruding from the beach, and those well above the beach (foreground) show identical erosion. Furthermore, the rock formations in the surf are jagged. These conditions would be impossible if, according to geology's accepted "no flood, ever" paradigm, the rocks in the ocean had been exposed to pounding surf for billions of years, because they would appear rounded or they would be eroded entirely. Note the gray color of the uppermost layer of topsoil on the right side of the image.

Figure 11 captures a particularly interesting feature found on the California coastline. The photograph was taken during low tide. What makes it so interesting is its central feature, a small rock peninsula topped by less than a meter of gray soil that is protected on the surface by invasive ice plant. This tiny peninsula is very close to the ocean during low tide and nearly directly above it during high tide; in addition, it is only 6 or 7 m above the high tide level. We can assert that the topsoil layer could not possibly have formed above the small peninsula if there were never a flood.

Figure 10. Photograph of jagged rocks along the California coastline showing identical erosion in the surf, exposed on the beach, and above the beach.

The isolated topsoil layer is easily explained according to the correct perspective that there was a world-wide flood: the layer was contiguous with and exactly similar to all the other topsoil in the nearby region (noticeable on Figure 10, just right of center), including those from lower-elevation rocks that are now submerged. The common soil layers formed over eons of varying weather, but especially from persistent rains that eroded the bedrock terrain features. The soil layer was mutually self-supporting until intrusion by floodwaters removed the layer from formations in or exposed to the water, thereby making adjacent and partially removed soil layers unstable. Unlike nearby features in and near the ocean waters, most of the tiny peninsula's upper layer of soil has somehow survived the 13,000 years of Post-Diluvian ocean and environmental activity. In addition, it has not been affected by tsunami activity.

Figure 11. Photograph of a small rock peninsula above a beach on the California coast. In the center of the photograph is a segment capped by a layer of topsoil (gray materials beneath non-indigenous ice plant) that has yet to be washed away like its neighboring sections.

The topsoil layer that once covered lower and now-submerged landscapes in the local region has been removed by the ocean waters, exposing the jagged rocks beneath, and the soil remnants now form portions of the area's ocean bed. Knowing that former soils have been mixed with other pre-flood soils to form littoral ocean sediments should help submarine geomorphologists to better analyze and understand the morphology of samples obtained from such regions.

The flood displaced the atmosphere upward, meaning that the landscapes we presently occupy would have been exposed to significantly less pre-flood atmospheric pressure because they were more than 3 km above the former sea level. An estimate obtained from models of atmospheric pressure versus altitude indicates that air pressure 3.5 km above sea level is roughly 60% of standard atmospheric pressure; landscapes roughly 5 km above sea level would have slightly less than 50% (see model and chart in Appendix B).

Interestingly, a recent publication authored under the prevailing "no flood, ever" paradigm analyzed gas bubbles obtained from basaltic lava flows in Australia that reportedly solidified several billion years before present at what its investigators take to be present sea level. Investigators concluded that atmospheric pressure was less than one half modern levels (Som et al. 2016).

Figure 12 presents a map of the Beasley River region in Australia where the study's researchers obtained their lava samples. Also shown is the bathymetry of a portion of the ocean to the west where, similar to the western Mediterranean Sea, the drainages in the former basin ended at a common terminal depth, in this case approximately 5.2 km below present sea level. Rather than forming at sea level, the lava instead formed more than 5 km above sea level where we expect the atmospheric pressure to be less than half of present. Thus, Earth's atmosphere has been relatively constant over the course of its history.

Such diminished atmospheric pressure in these formerly upland regions likely accounts for the evolution of large flightless birds that we encounter, post-flood. Before the flood, and at 50% of current atmospheric pressure, these birds could not develop the requisite lift to attain flight.

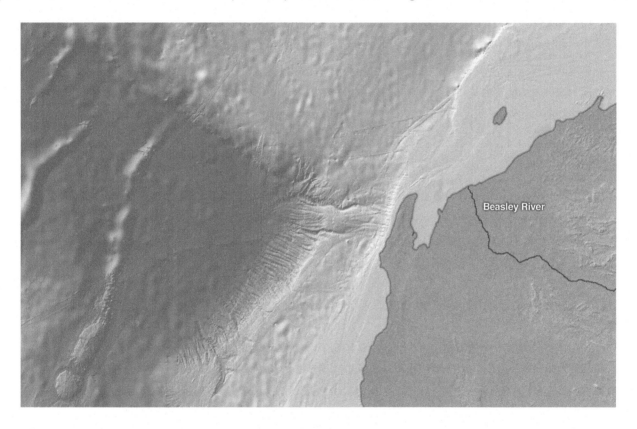

Figure 12. On this map are shown the coast of northwest Australia, through which the Beasley River flows, as well as bathymetry details to the west that show the terminal depth common to the many former tributaries.

Another recently published discovery, that of a massive subglacial trough 300 km long, up to 25 km across, deeper than the Grand Canyon, and more than 2 km below present sea level (Ross, 2013), is easily understood in the context of the worldwide flood. Because ice floats on water, glacial scouring 2 km beneath present sea level could never have occurred had the present amount of water always existed, as currently assumed. However, in the context of the worldwide flood, the glaciers formed in Antarctica flowed down into the former abyss, subaerially scouring the valley over the eons before the flood. They have since been covered and preserved by the flood waters.

Chapter 7
Resolving the Problems of Atlantis

We would seem to be left with only two alternatives: (1) to presume exaggeration on the part of one of the transmitters of the tale and situate a vastly shrunken Atlantis somewhere other than the Atlantic Ocean, or (2) presume outright invention and discard the idea of historical Atlantis altogether. . . . I would offer a third choice: withhold judgment. . . . It may be a change in our perception of what is, or was, possible that is wanted here. At the very least our perceptual field must be widened.

Dr. Mary Settegast (1987)

Details of ocean bathymetry have widened our perceptual field so much that they overturn an erroneous scientific paradigm; they afford a better perception of what once was. Yet in addition to revealing the cause of the worldwide flood, the maps allow us to resolve the problem of Atlantis. We can now pass judgment: Atlantis was lost to the worldwide flood.

In *Critias*, one of Plato's dialogues, he describes the Atlantis canal system, as follows (italics added):

It was rectangular, and for the most part *straight and oblong*. . . . It was excavated to the *depth of a hundred feet, and its breadth was a stadium [equivalent to 185 meters] everywhere; it was carried round the whole of the plain, and was ten thousand stadia in length. . . . The depth and width and length of this ditch were incredible* and gave the impression that such a work, in addition to so many other works, could hardly have been wrought by the hand of man. It received the streams which came down from the mountains, and *winding round the plain*, and touching the city at various points, was there *left off into the sea*. . . . From above, likewise, straight canals of a hundred feet in width were cut in the plain, and again let off into the ditch toward the sea; these *canals were at intervals of a hundred stadia, . . . cutting transverse passages from one canal into another*, and to the city.

Figure 13 is a NOAA map, centered at 24.4°W, 31.3°N, that shows the Atlantis canal system. The map allows us to compare the canals with Plato's description. First, we note that the canals were straight and formed rectangular sections. The canal perimeter measures approximately 165 km east to west and 120 km north to south, so it was immense, which leads one to wonder how long it must have taken to build. In addition, its canals were sufficiently deep and wide to be discerned by modern instruments. The canals' water source was likely the highland region in the west, which is the eastern extent of the Mid-Atlantic Ridge. We can see that the interlocking transverse canals were mostly at right angles and that the system might have drained to the northeast, where we find the drainage ditch's channel. The distance between the canals varies, but the span between two major east-west canals, identified by the red arrows on Figure 13, measures 15 km, which equates to approximately 85 stadia (assuming that 5.666 stadia equal 1 km). Thus Plato's description of the distance between canals is close to what we observe, though only somewhat exaggerated.

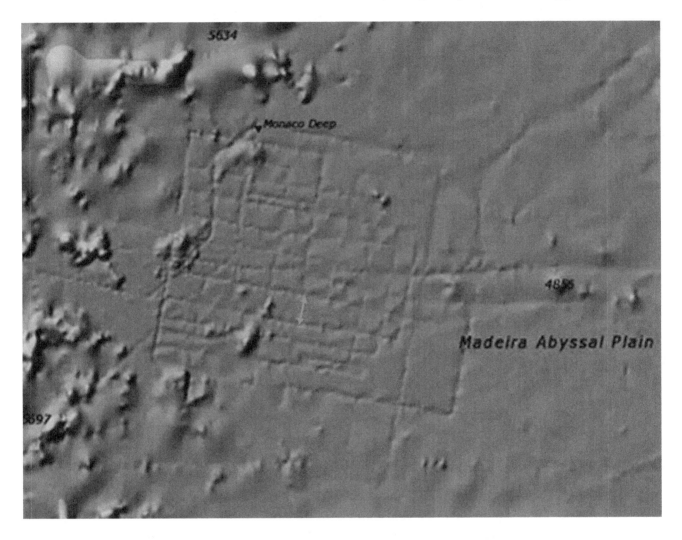

Figure 13. The canals of Atlantis are found in the Madeira Abyssal Plain. The center of the canal system is located near 24.4°W, 31.3°N. The red arrow measures 15 km or about 85 stadia (a bit less than 100 stadia as described by Plato). Image source: NOAA.

To determine the overall length of the canals, we can overlay straight line segments on the grid as shown on Figure 14. Then we can take those segments, lay them end to end, and convert their distance in kilometers to stadia. The length of the canal system is calculated to be 1,775 km, which translates to nearly 9,600 stadia, a number within 4% of Plato's description.

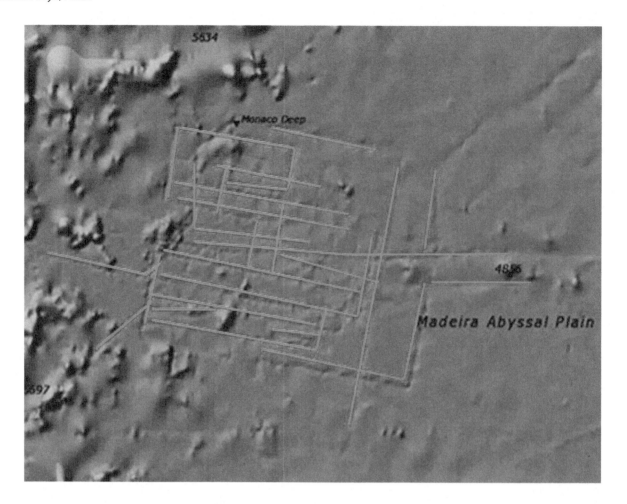

Figure 14. By overlaying straight lines on the canals of Atlantis, we can approximate its total length in kilometers, then convert that to stadia. Source: NOAA.

Figure 15 shows that the canals are approximately 1,750 km west-southwest of the Strait of Gibraltar near the Canary Islands, 750 km south of the Azores, and 650 km nearly due west of Madeira. Atlantis existed in the location that Dr. Settegast and other prehistorians anticipated.

Atlantis' fate is described in Plato's *Timaeus* (italics added):

> At a later time *there were earthquakes and floods of extraordinary violence*, and in a single dreadful day and night all your fighting men were swallowed up by the earth, *and the island of Atlantis was similarly swallowed up by the sea and vanished*.

The incredible earthquakes that Plato recounts would have been induced by the immense cosmic impact. Soon thereafter the newly introduced floodwaters coursed their way around the planet from the impact area and into low-lying regions such as the Madeira Abyssal Plain where Atlantis existed.

Plato's description, coupled with the new map data, allows us to resolve the problem of Atlantis. It was buried by the worldwide floodwaters.

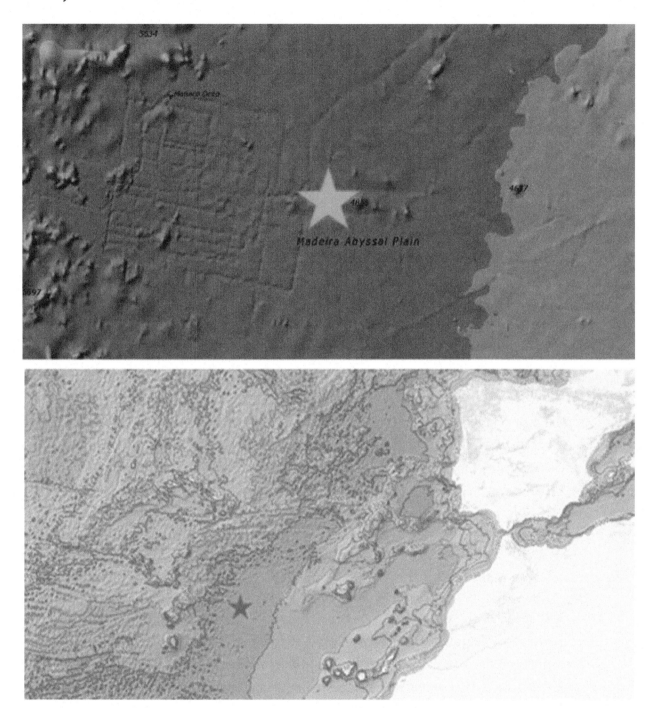

Figure 15. The blue star in each map indicates the location of the Atlantis canal system. Source: NOAA.

Chapter 8
Lemuria (Mu)

We are forced to travel through distant lands, and become familiar with the complexions, and the feelings, and the characters of mankind under every form of social life. . . .

Adam Sedgwick (1831)

Not many of us are aware of Lemuria (also known as Mu), an ancient civilization that some claim existed in a great expanse of the tropics now covered by the Pacific Ocean. As a consequence of geology's accepted "no flood, ever" doctrine, Lemuria's existence has been castigated as the musings of eccentrics (at best). But because geologists erred, we can now reconsider the legend of Lemuria from a more informed perspective. Similar to the legend of Atlantis, Lemuria's history and fate are worthy of discussion.

Lemuria comes to our attention mainly through the 1926 publication of James Churchward's *The Lost Continent of Mu: Motherland of Men*, which resulted from his translation of ancient tablets he discovered in India. His book culminates 50 years of research, and it prompted Churchward to claim that "at one time the earth had an incalculably ancient civilization which was, in many respects, superior to our own, and far in advance of us in some important essentials that the modern world is just beginning to have the cognizance of."

Churchward describes Mu (italics added):

> The civilizations of the early Greeks, the Chaldeans, the Babylonians, the Persians, the Egyptians and the Hindus had been definitely preceded by the civilization of Mu. Continuing my searches, *I discovered that this lost continent had extended from somewhere north of Hawaii to the south as far as the Fijis and Easter Island*, and was undoubtedly the original habitat of man. I learned that in this beautiful country there had lived a people that colonized the earth, and that *this land of smiling and plenty had been obliterated by terrific earthquakes and submersion 12,000 years ago, and had vanished in a vortex of fire and water. . . .*

> It was a beautiful tropical country with *vast plains*. The valleys and plains were covered with rich grazing grasses and tilled fields, while low rolling hill-lands were shaded by luxuriant growths of tropical vegetation. No mountains or mountain ranges stretched themselves through this earthly paradise. . . .

At the time of our narrative, *the 64,000,000 people were made up of ten tribes or peoples, each one distinct from the other, but all under one government....*

Rumblings from the bowels of the earth, followed by earthquakes and volcanic outbursts, shook up the southern part of the land of Mu.... During the night the land was torn asunder and rent to pieces. *With thunderous roarings the doomed land sank.* Down, down, down she went, into the mouth of hell – a tank of fire.... As Mu sank into that gulf of fire another force claimed her – fifty million square miles of water. *From all sides huge waves or walls of water came rolling in over her.* They met where once was the center of the land. Here they seethed and boiled.

Poor Mu, the motherland of man, with all her proud cities, temples and palaces, with all her arts, sciences and learning, was now a dream of the past. *The deathly blanket of water was her burial shroud. In this manner was the continent of Mu destroyed.... For nearly 13,000 years* the destruction of this great civilization cast a heavy pall of darkness over the greatest part of the earth.

Both Plato and Churchward claim that ancient civilizations, Atlantis and Mu, respectively, were lost 12,000 years ago after great earthquakes were followed by submersion in a flood. Although it is possible that Churchward's writings were influenced by Plato, if the two accounts are independent, then their similarity presents a compelling coincidence.

Churchward was a talented painter, and he complemented his archaeological investigations with many works. Several of his paintings depict the destruction of Mu, and a portion of one painting's caption reads "Unaccompanied with Ice," which I find fascinating. The celestial object that delivered the floodwaters was composed mostly of ice, and not all of it would have melted immediately after impact – some of it would have been carried about by the relentlessly coursing waters. Because Mu was located relatively close to the impact site, it is very reasonable to consider that the meltwaters would have carried substantial amounts of ice with them. Yet Churchward was hamstrung by geology's "no flood, ever" dogma; because he postulated that the entire continent of Mu had subsided into the abyss, he would not have considered that the waters had a frozen, cosmic source. Yet what prompted him to mention ice? Where would he have found the account? Is it available today?

It is worth noting that another of Churchward's paintings of Mu's destruction depicts volcanic activity, which is also a likely and immediate consequence of the cosmic impact.

In addition to paintings, Churchward also created maps of Mu that he derived from his research. Its borders bear remarkable similarity to the expanse of formerly exposed landscapes in tropical Pacific latitudes to the west of the Americas, shown in tan on the ArcGIS map pre-flood model on Figure 7.

Further support of Lemuria's pre-flood existence comes from a recent paper by Llamas et al. (2016) dealing with DNA similarities between indigenous people in Australia and in the Amazon in South America. The DNA heat map from the Llamas et al. paper is shown on Figure 16.

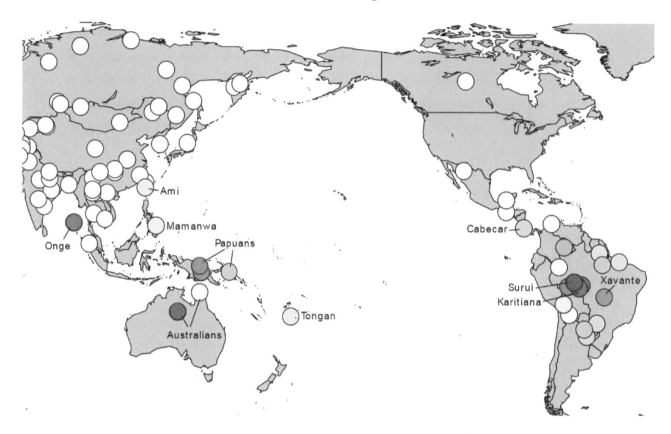

Figure 16. DNA heat map showing inferred closest similarities between indigenous humans from Australia and South America (deep red) and lesser similarities in lighter colors. White circles indicate very few inferred DNA similarities. (Image reproduced with permission from Smithsonian.)

Explanation of the heat map is obvious in the context of the worldwide flood. As described by Churchward, the peoples inhabiting Mu shared a common ancestry, and descendants of flood survivors in Australia and regional Pacific islands remain DNA-linked to descendants of flood survivors in South America. As implied by the DNA heat map, the existence of Lemuria and its loss to the worldwide flood corroborate that South America was not populated by way of North America.

Like Atlantis, Lemuria (Mu) became submerged in the worldwide floodwaters introduced by the cosmic impact approximately 13,000 years before present.

Chapter 9
Conclusions and Implications

The earth is flat.
The earth is the center of the universe.
There was never a worldwide flood.

Significant errors throughout the history of science

A cosmic impact nearly 13,000 years before present introduced more than 3 km of water to the earth's ocean basins and ecosystem. By causing the submersion of vast formerly exposed landscapes, by displacing the atmosphere, and by inducing what we call the Younger Dryas ecosystem changes, the worldwide flood forever changed the planet. The impact ushered in a new geologic era, the Post-Diluvian.

Geology's prevailing "no flood, ever" paradigm is the most profound error in the history of science. It rests on an indisputable error: by concluding that there was never a worldwide flood, early geologists precluded the possibility that submerged landscapes might at one time have been exposed and later flooded. They assumed that the present amount of water has always been with the planet, and in so doing they passed judgment on landscapes that they could not observe. As a consequence, an entire branch of science has been fundamentally wrong for nearly 200 years. Geology wholly and completely misunderstands the morphology of both exposed and submerged landforms. Only by accident might submarine geomorphology have ever correctly analyzed sediments derived from oceans and seas that existed before the flood. That a major branch of science could be so wrong for so long is most remarkable, particularly at this stage of human understanding. Until geology reforms, its existing findings and publications must be treated as pseudo-science, especially its most egregious violation of the scientific method: the fantasy that ascribes turbidity flows as the creation mechanism for submerged structures such as Monterey Canyon.

The observations afforded by Google Maps and Google Earth and the resulting paradigm shift are historically equivalent to Galileo's use of the telescope that led to overturning a prevailing yet incorrect belief: geocentrism. Ultimately, every scientific discipline must accept and incorporate the correct paradigm, namely, that there was a worldwide flood.

Culturally independent, ubiquitous accounts of a worldwide flood that have been discounted or dismissed due to geology's fundamental error are instead factual, corroborating accounts by survivors. Ensuing, significant cultural changes induced by the worldwide flood led the Göbekli Tepe to commemorate the event in stone carvings.

The serpent-like appearance of the IO and its debris tail as they neared Earth and its commemoration as Satan (in one culture) render the biblical account of Adam and Eve as but another flood story. To wit: abrupt, serpent-like-IO/Satan-induced environmental changes forced furless (naked) humans out of the habitat for which they were adapted (Eden), and this fundamentally altered their nature as they must now struggle to survive.

Recognizing that there was a worldwide flood will revolutionize anthropology. For instance, humans evolved in formerly exposed abyssal landscapes; we are not out of Africa. Increased atmospheric pressure caused warm temperatures in the pre-flood abyss; this explains why humans do not have fur. (Correspondingly, our simian relatives have fur because they adapted to landscapes that, at more than 3 km above the former sea level, were much colder.) The worldwide flood accounts for the seemingly sudden and/or inexplicable appearance of humans in presently exposed landscapes. Human flood survivors either lived in tropical upland regions before the event, or they made their way to higher elevations and habitable landscapes during the event. Because furless humans evolved in the warm abyss, we find ourselves maladapted to the cooler Post-Diluvian ecosystem. This explains our need for clothing, warmth, and shelter. Within this context, it is evident that *everything* we have done and will continue to do, post-flood, is for the purpose of our survival.

With the exception of some tropical latitudes that bear evidence of pre-flood human activity, formerly upland regions were mostly or completely uninhabited by humans because they were too cold, especially during glacial periods. At least two civilizations that we know of, Atlantis and Lemuria, were lost to the worldwide flood. Remnant canals from Atlantis are identifiable in maps of the Madeira Abyssal Plain, and Plato's description of the canal system presents a compellingly close match to an analysis of the region. Atlantis's artifacts can and will be discovered. Lemuria (Mu) also existed across a vast expanse of tropical landscapes under what is now the Pacific Ocean west of the Americas; to this day its descendants in Australia and South America share a very close DNA link.

Published human population estimates applied to exponential growth models yield a wide range in the number of flood survivors. However, that number is in the thousands, from a low of 2,000 to a high of 400,000. (The disparity is due to the variance in published population estimates for the years 1800 CE and 0 CE that affect important parameters in the mathematical model for population growth. In each case, the model, discussed in Appendix B, assumes constant population growth rates for the periods from year 10800 BCE to 1800 CE.)

Earth's temperature record derived from polar ice cores is flawed because it is based on the assumption that the poles and the atmosphere above them have always been as they are now. Instead, at more than 3 km above the former sea level, the atmosphere in which the majority of the ice formed was far different from that of the last 13,000 years. Because this temperature record is flawed, it cannot be used for science or policymaking. In order for it to be employed for such purposes, it must be revised to account for the flood's effects.

Before we enact policy decisions on climate change, perhaps first we should understand ecological changes that have transpired during the last 13,000 years and address such questions as the following: Has the planet's ecosystem attained some sort of equilibrium? Will the waters modulate the planet's temperature to preclude future ice ages? Furthermore, in the context of our maladaptation to the post-flood Earth ecosystem, we also should address: what sociopolitical and economic system(s) and/or idea(s) might best ensure our survival? Is such a thing even possible? Perhaps we might ponder: is our abuse of the environment a natural consequence of a sentient yet maladapted species seeking to survive?

Finally, among the root causes of our most visceral differences is the divide created by the "no flood, ever" paradigm, for geologists initiated the "science" vs "religion" schism from which followed a condescension toward those who follow human narrative traditions. Let us find some kernel of unity in the realizations that there was a flood and that we are but an ill-adapted species trying to survive in Post-Diluvian Earth.

Afterword

We are justifiably within bounds to wonder whether there are other theories that geologists might have wrong. For instance, they have us believe that swirling currents inside the planet somehow reach up and slowly move the continents. How in the world does that happen? What permissible modes of circulation inside a sphere (a problem analogous to the vibrations of a guitar string) would lead to such asymmetries as those observed in Earth's landform distribution? Would geologists' assumed circulatory patterns not violate mathematics? What of Newton's third law of motion (for every action there is an equal and opposite reaction)? Do we suspend it for tectonic movements?

Rhetorical questions, yet again.

The view of Earth shown on Figure 17 contains sufficient information to explain landform configuration as well as the initiation of the planet's obliquity (tilt). We will address this matter later. Until then, please consider that the following map is centered on South America intentionally.

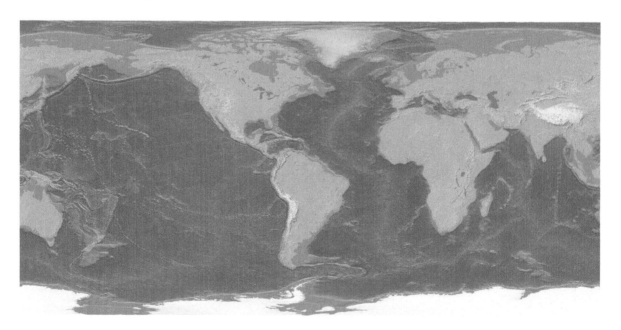

Figure 17. Bathymetry details on this world map reveal impact trough remnants on the northern and southern extents of South America. Associated landmass movement traces are evident beneath the oceans. Impact sites are nearly half the world west of the trough remnants.

Appendix A Excerpt from Adam Sedgwick's Farewell Address to the Geological Society of London

In retreating where we have advanced too far, there is neither compromise of dignity nor loss of strength; for in doing this we partake but of the common fortune of every one who enters on a field of investigation like our own. All the noble generalizations of Cuvier, and all the beautiful discoveries of Buckland, as far as they are the results of fair induction, will ever remain unshaken by the progress of discovery. It is only to theoretical opinions that my remarks have any application.

Different formations of solid rock, however elevated and contorted, can never become entirely mixed together; and the very progress of degradation commonly lays bare all the elements of their structure. But diluvial gravel may be shot off from the flanks of a mountain chain, during one period of elevation, and become so confounded with the detritus of another period, that no power on earth can separate them; and every subsequent movement, whether produced by land floods or any other similar cause, must continually tend still further to mingle and confound them. The study of diluvial gravel is, then, not only one of great interest, but of peculiar difficulty and nice discrimination; and in the very same deposit, we may find the remains of animals which have lived during different epochs in the history of the earth.

Bearing upon this difficult question, there is, I think, one great negative conclusion now incontestably established – that the vast masses of diluvial gravel, scattered almost over the surface of the earth, do not belong to one violent and transitory period. It was indeed a most unwarranted conclusion, when we assumed the contemporaneity of all the superficial gravel on the earth. We saw the clearest traces of diluvial action, and we had, in our sacred histories, the record of a general deluge. On this double testimony it was, that we gave a unity to a vast succession of phaenomena, not one of which we perfectly comprehended, and under the name diluvium, classed them all together.

To seek the light of physical truth by reasoning of this kind, is, in the language of Bacon, to seek the living among the dead, and will ever end in erroneous induction. Our errors were, however, natural, and of the same kind which led many excellent observers of a former century to refer all the secondary formations of geology to the Noachian deluge. Having been myself a believer, and, to the best of my power, a propagator of what I now regard as a philosophic heresy, and having more than once been quoted for opinions I do not now maintain, I think it right, as one of my last acts before I quit as Chair, thus publicly to read my recantation.

We ought, indeed, to have paused before we first adopted the diluvian theory, and referred all our old superficial gravel to the action of the Mosaic flood. For of man, and the works of his hands, we have not yet found a single trace among the remnants of a former world entombed by these ancient deposits. In classing together distant unknown formations under one name; in giving them a simultaneous origin, and in determining their date, not by the organic remains we had discovered, but by those we expected hypothetically hereafter to discover, in them; we have given one more example of the passion with which the mind fastens upon general conclusions, and of the readiness with which it leaves the consideration of unconnected truths.

Are then the facts of our science opposed to the sacred records, and do we deny the reality of a historic deluge? I utterly reject such an inference. Moral and physical truth may partake of a common essence, but as far as we are concerned, their foundations are independent, and have not one common element. And in the narrations of a great fatal catastrophe, handed down to us, not in our sacred books only, but in the traditions of all nations, there is not a word to justify us in looking to any mere physical monuments as the intelligible records of that event: such monuments, at least, have not yet been found, and it is not perhaps intended that they ever should be found. If, however, we should hereafter discover the skeletons of ancient tribes, and the works of ancient art buried in the superficial detritus of any large region of the earth; then, and not till then, we may speculate about their stature and their manners and their numbers as we now speculate among the disinterred ruins of an ancient city.

We might, I think, rest content with such a general answer as this. But we may advance one step further – History is a continued record of passions and events unconnected with the enduring laws of mere material agents – The progress of physical induction on the contrary, leads us on to discoveries, of which the mere light of history would not indicate a single trace. But the facts recorded in history may sometimes, without confounding the nature of moral and physical truth, be brought into a general accordance with the known phaenomenon of nature; and such general accordance I affirm there is between our historical traditions and the phaenomenon of geology. Both tell us in a language easily understood, though written in far different characters, that man is a recent sojourn on the surface of the earth. Again, though we have not yet found the certain traces of any great diluvian catastrophe which we can affirm to be within the human period; we have, at least, shown, that paroxysms of internal energy, accompanied by the elevation of mountain chains, and followed by might waves desolating whole regions of the earth, were a part of the mechanism of nature. And what has happened, again and again, from the most ancient, up to the most modern periods in the natural history of the earth, may have happened once during the few thousand years that man has been living on its surface. We have, therefore, taken away all anterior incredibility from the fact of a recent deluge; and we have prepared the mind, doubting about the truth of things of which it knows not either the origin or the end, for the adoption of this fact on the weight of historic testimony.

If, Gentlemen, I believed that the imagination, the feelings, the active intellectual powers bearing on the business of life, and the highest capacities of our nature, were blunted or impaired by the study of our science, I should then regard it as little better than moral sepulcher, in which, like the strong man, we were burying ourselves and those around us, in ruins of our own creating. But I believe too firmly in the immutable attributes of that Being, in whom all truth, of whatever kind, finds its proper resting place, to think that the principles of physical and moral truth can ever be in lasting collusion. And as all the branches of physical science are but different modifications of a few simple laws, and are bound together by the intervention of common objects and common principles; so also there are links (less visible, indeed, but no less real) by which they are also bound to the most elevated moral speculations.

At every step we take in physics, we show a capacity and an appetency for abstract general truth; and in describing material things, we speak of them, not as accidents, but as phaenomena under the government of laws. The very language we use (and it is hardly possible for us to explain our meaning by any other), is the language in which we describe the operations of intelligence and power. And hence we admit, by the very constitution of our intellectual nature, and even in spite of ourselves, an *anima mundi* pervading all space existing in all times, and under all conditions of being.

But we do not stop here; for the moment we pass on to that portion of matter, which is subservient to the functions of life, we there find all the phaenomena of organization; and in all those being the functions of which we comprehend, we see traces of structure in many parts as mechanical as the works of our own hands, and, so far, differing from them only in complexity and perfection; and we see all this subservient to an end, and that end accomplished. Hence, we are compelled to regard the *anima mundi* no longer as a uniform and quiescent intelligence, but as active and anticipating intelligence; and it is from this first principle of final causes, that we start with that grand and cumulative argument, derived from all the complex functions of organic nature.

Geology lends a great and unexpected aid to the doctrine of final causes; for it has not merely added to the cumulative argument, by the supply of new and striking instances, of mechanical structure adjusted to a purpose and that purpose accomplished; but it has also proved that the same pervading active principle, manifesting its power in our times, has also manifested its power in times long anterior to the records of our existence.

But after all, some men seeing nothing but uniformity and continuity in the works of nature, have still contended (with what I think is mistaken zeal for the honour of sacred truth), that the argument from final causes proves nothing more than a quiescent intelligence, I feel not the force of this objection. In geology, however, we can meet it by another direct argument; for we not only find in our formations organs mechanically constructed – but at different epochs in the history of earth we have great changes of external conditions, and corresponding changes of organic structure; and all this without the shadow of proof that one system of things graduates into, or is the necessary and efficient cause of, the other. Yet in all these instances of change, the organs, as far as we can comprehend their use, are exactly those which were best suited to the functions of the being. Hence we not only show intelligence contriving means adapted to an end, but at successive times and periods contriving a change of mechanism adapted to a change in external conditions. If this be not the operation of a prospective and active intelligence, where are we to look for it?

Our science is then connected with the loftiest of moral speculations; and I know no topic more fitting to the last sentiments I wish to utter from this Chair.

There is one way, and one way only, in which the higher intellectual powers may be cramped by the pursuit of natural truth, and that is by a too exclusive devotion to it. In the pursuit of any subject, however lofty, a man may become narrow-minded, and in a condition little better than that of a moral servitude: but on this score we have not much to fear. Every department of science offers its spoils for our decoration; we are carried into regions where we contemplate the most glorious workmanship of Nature, and where the dullest imagination becomes excited; we are forced to travel through distant lands, and become familiar with the complexions, and the feelings, and the characters of mankind under every form of social life; and in doing this, if we be not most indocile learners, we must bear away lessons of kindness, and forbearance, and freedom of thought, along with the appropriate knowledge of our own vocation; and all this we can carry with us into the business of life. These, Gentlemen, are the high qualities which ought to form the ornament of this Society; and I am certain that I have seen their constant exercise in the intercourse and the discussions of this room, where mutual goodwill, frankness, and the love of truth, are the only dominant sentiments.

My own connexion with this Society during the two years I have had the honour to preside over its councils, has been to me a source of continued and heartfelt pleasure; and it would be with pain indescribable that I should now quit this Chair and bid you farewell, did I not think that I should very often meet the same friends, and partake in the same discussions.

Every man, whatever be his station, has a small circle of duties which are paramount to all others; but after these are performed, such powers as are given me shall ever be willingly devoted to your service. I do not mean this for empty boasting; that language would ill become me at any time, and least of all when I am leaving this Chair and descending into your ranks. Mine has been indeed but an interrupted service; but I resign it to one of whose powers you have had long experience, who can give them to you undivided, and whose hands are in no respect less ready than my own."

Appendix B Supporting Materials – IO Ice Fragment Analysis, IO Water Volume, Population Model, Atmospheric Pressure, Geometry

1. The fragile IO: its ice fragments rained down on North America and South America immediately prior to impact

IO fragility can be inferred from an understanding of the small aggregating accelerations induced by its central core. We apply Newton's law of gravitation and his second law of motion to determine accelerations induced by one mass on another, which can be expressed as follows:

$a = G * M/r^2$

where: M is the mass of the attracting object, r is the distance to that object's center, and G is the universal gravitational constant, $G = 6.67 * 10^{-11}$.

From this equation, we can show that the acceleration of an object near the earth's surface is roughly 9.8 m/s² (first set of calculations, below). Then we compute the acceleration at the surface of a 50 km sphere composed of very dense material. Afterward, we add a porous ice-debris outer layer like that of the IO in order to calculate the acceleration at its outer surface.

Acceleration at Earth's surface:

Radius:	6,380,000	meters	
Volume:	1.0878E+21	meter^3	
Density:	5497.31393	kg/meter^3	
Mass:	5.98E+24	kg	
Acceleration at surface:	**9.799088059**	**m/sec^2**	=G*mass/radius^2

Acceleration at the IO's core surface:

Radius:	50,000	meters
Volume:	5.23599E+14	meter^3
Density:	5497.31393	kg/meter^3
Mass:	2.87839E+18	kg
Acceleration at core surface:	**0.076795361**	m/sec^2 =G*mass/radius^2

Acceleration at the IO's outer surface:

Radius:	1,250,000	meters
Volume:	5.23599E+14	meter^3
mass, 1km^3 water:	1E+12	kg
H20 mass, outer layer:	1.28779E+21	kg
mineral mass, outer layer:	5497.31393	kg
Acceleration due to core + shell masses:	**0.198841398**	**m/sec^2** =G*[(H20+mineral mass, outer layer) + core mass]/outerradius^2

IO's acceleration, expressed as a fraction of Earth's acceleration:

a_core/a_earthsurface	= 0.007836991	= 0.70%
a_combined mass core + shell/a_earthsurface	= 0.020291827	= 2%

Thus, the IO induced very small attracting accelerations, which is the primary cause of its porous nature.

The gap in the impact crescent (e.g. Figure 5) indicates that the IO was falling apart as it neared impact, an observation supported by an abundance of impact craters strewn along the IO's broad and lengthy approach path. We can get an idea of the IO's approach from the impact trough that was carved by its dense nucleus: back-propagating the trough direction reveals the impact approach path. In doing so we find that the IO's center of mass approached over west-central North America then western South America before crossing Chile and Argentina and flying over the Falklands. Its approximate path is depicted on Figure B1.

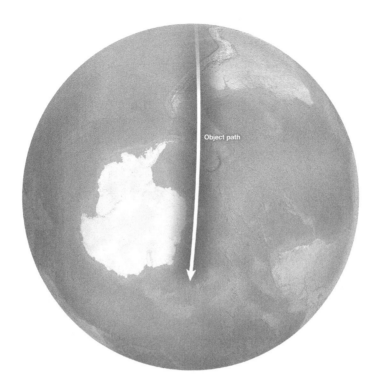

Figure B1. The arrow approximates the path taken by the IO's center of mass.

IO fragments would pass through less atmosphere as it traveled southward: the IO was outside the atmosphere when over North America, and it was much lower as it traveled over the southern part of South America. The IO's decreasing altitude means that there would be lesser atmospheric heating and ablation experienced by fragments falling along the southern parts of the approach. Longer atmospheric exposure during fall could affect crater shape, meaning that North American craters might be somewhat different from those found in South America. In addition, the nature of the landscape upon which the ice fragments fell would have affected crater shapes; however, and unfortunately for us, the presently exposed landscapes upon which we observe the ice impact craters was more than two miles above the pre-flood sea level, and, as a consequence, we cannot be certain of their pre-flood constitution. Nonetheless, the new maps provide an opportunity to discover and investigate craters caused by the IO's fragments.

Some of the more famous North American impact craters are the elliptical Carolina Bays. A topic of debate among geologists (mentioned below), the bays are among many ice impact craters found in parts of North America. Due to the IO's path, we expect to find craters in South America, too, something that Google Earth allows us to accomplish – they are found along the entire length of the continent (admittedly, the Amazon jungle makes map identification there nearly impossible). We should note that the South American impacts have received little or no mention in the geologic literature, something attributable to one of two reasons: either geologists are unaware of the myriad South American impact craters, or they cannot fit them into their North American YD impact hypothesis.

It turns out that South America has so many craters that the number dwarfs that from North America. The increased number is likely due to more intense atmosphere-induced fragmentation on the IO as its altitude decreased over the continent. The craters' orientations, largely from north-northwest toward the south-southeast, precludes an Antarctic impact as a possible source. A swarm of impact craters from south-central Argentina is shown on Figure B2.

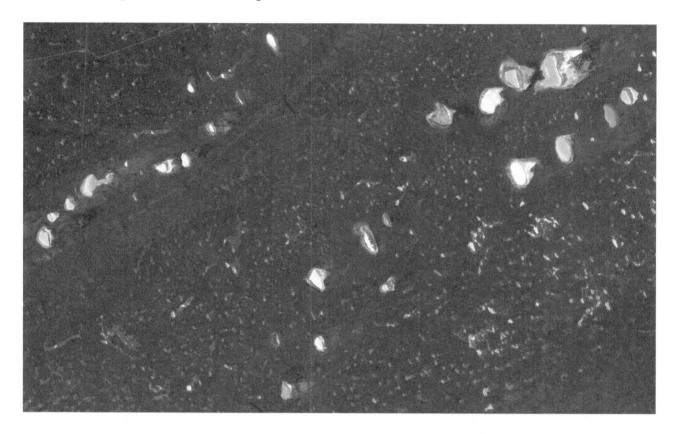

Figure B2. Several hundred IO fragment-created craters of various sizes are shown in this map. The long axes of the larger craters measure approximately one kilometer whereas the smaller craters are one-tenth that size.

Listed on the table, below, are latitude-longitude coordinates for some ice fragment impact craters along the IO approach path. The list is not intended to be exhaustive; rather, it is meant to illustrate the multitude of craters created by IO ejecta. The latitude-longitude pairs are provided so that you might discover them – and others – using Google Earth. The term "eye" refers to the suggested altitude from which to begin crater investigation. The "Comments" describe locations and some features intended to pique interest. The progression of impacts listed on the table moves the viewer from north to south.

Impact Coordinates	Eye		Comments
Northern Latitudes			
40.6341N 98.0162W	7,800	ft	Nebraska; might be difficult to discern among crop circles
40.4670N 98.0381W	17,000	ft	Nebraska
39.1658N 75.8462W	3,500	ft	Maryland
34.8719N 79.0371W	46,600	ft	South Carolina, swarm of elliptical craters
34.8370N 79.1854W	20.3	mi	South Carolina, elliptical craters
32.8604N 82.0342W	12.5	mi	Georgia

33.4013N 104.0641W	40,600	ft	New Mexico
34.6756N 103.9874W	37,500	ft	New Mexico, swarm
34.8448N 104.1021W	45,000	ft	New Mexico, swarm
32.2140N 102.4217W	30,600	ft	Texas, swarm with one crater in someone's backyard
32.5304N 100.6679W	17.5	mi	Texas
26.3530N 97.7112W	28,300	ft	Mexico
25.7206N 97.3893W	23.7	mi	Mexico
20.3999N 87.4530W	37,800	ft	Mexico; impact string visible at large view scale, running SW-NE
20.0234N 87.5228W	40	mi	Mexico, swarm of large impact craters; some carved shoreline
19.1279N 87.8039W	16	mi	Mexico, swarm
18.3340N 88.2799W	13	mi	Mexico
14.4011N 83.3440W	13,000	ft	Mexico

Southern Latitudes

6.1710S 80.7380W	28,500	ft	Peru; equatorial latitude impact crater
10.6985S 76.3237W	28.0	mi	Peru; two elongated impacts in mountainous region
22.8193S 66.8091W	47,000	ft	Argentina; swarm
34.8117S 61.6309W	20	mi	Argentina
35.0281S 62.4160W	31,000	ft	Argentina
35.8648S 62.3402W	32,000	ft	Argentina; impact swarm
37.4198S 58.1596W	16,000	ft	Argentina
37.6990S 61.0177W	18	mi	Argentina; swarm
41.2603S 68.0857W	13.5	mi	Argentina; check out the drainage runoff patterns from ice melt
41.3549S 67.7267W	16	mi	Argentina; swarm
45.1512S 70.6540W	24	mi	Argentina; small swarm, some drainages observable
47.8645S 71.4748W	25	mi	Argentina; impact crater now a lake; swarm in vicinity
50.5908S 70.3878W	37	mi	Argentina; large swarm
51.5756S 70.0404W	30	mi	Argentina; large swarm
51.9179S 70.0099W	30	mi	Argentina (barely); large swarm
51.7803S 59.1534W	28,000	ft	Falkland Islands
53.6401S 68.2996W	40	mi	Argentina; swarm of large craters
55.8726S 67.8791W	7000	ft	Tierra del Fuego; swarm

[Geologists debate the formation of the Carolina Bays in North America. For instance, and under the prevailing "no flood, ever" paradigm, a recent publication ascribes the bays' creation – as well as the creation of all the North American craters – to some hypothetical impact on the Laurentide Ice Sheet (Zamora 2017). This presumed North American impact is said to have projected huge ice chunks from somewhere in the Great Lakes region to outside the atmosphere and then to where we find the craters several hundred to more than a thousand miles distant. Such an energetic impact should be easy to identify, no? Well, geologists have yet to identify it, and they never will, for it does not exist in North America.

In contrast, another publication claims that impact mineral markers indicate that the bays were created by "a previous unobserved, possibly extrasolar body" and suggests that a "comet fragmented and exploded over the Laurentide Ice Sheet creating numerous craters" (Firestone, 2009). The author back-propagated impact direction lines drawn from North American craters, and their many intersections represented candidate locations for supposedly projecting impacts. Unknown to the author, the set of intersections represents the approximate path taken by the IO.

Yet the geologists' debate does not matter, for it is another example of pseudo-science formed under an incorrect paradigm. Under the proper context that there was a worldwide flood delivered by the cosmic impact in what is now the Southern Ocean: all the craters in North America **and** South America were created by ice and other fragments that rained down along the IO's approach path.]

Therefore, the IO's small attracting accelerations in the Oort Cloud created a porous and fragile object that began to fall apart as it neared Earth impact. Furthermore, IO fragility led to the gap in its impact crescent (through which ice and other minerals were strewn great distances due to impact velocities), as well as to the ice craters in North America and South America.

2. Estimating the volume of water delivered by the impacting object:

We can estimate the volume of water delivered by IO by using the equation $V = 4/3*pi\ r^3$, where r is the diameter of the sphere and π (pi) is approximately 3.14159. . . . Since the impact diameter measures 2,500 km, then its radius measures 1,250 km.

IO radius:	1,250	km	
IO volume:	8,181,230,869	km^3	
Ice percentage:	17	%	=(1/4 non−open space)*2/3 non− open is ice
IO ice volume:	1,390,809,248	km^3	
IO water volume:	1,287,786,340	km^3	=ice vol/1.08
Ocean surface area:	361,000,000	km^2	
Ocean depth due to IO:	3.567275181	km	

Here we have found the volume of water delivered by the object and then divided it by the surface area that the oceans now occupy. Since the volume divided by the surface area yields the average depth of water, then this impact delivered roughly 3.57 km to the former abyss.

3. Population model, assumes change is proportional to its current size:

dP/dt = r P(t)

where: $P_{est1}(1800) = 900,000,000$ and $P_{est2}(1800) = 990,000,000$

and $P_{est1}(0) = 150,000,000$ and $P_{est2}(0) = 300,000,000$

then various combinations of those estimates yield high and low population estimates:

$P_{low}(-10800) = 1815$

and

$P_{high}(-10800) = 232294$

4. Atmospheric pressure model:

$P_{atm} = p_0 * \exp(-h/h_0)$, where $h_0 = 7$ km and $p_0 = 1$ atm. Plotting $P_{atm} = 1 * \exp(-h/7)$ for h between 0 and 9000 m yields this curve:

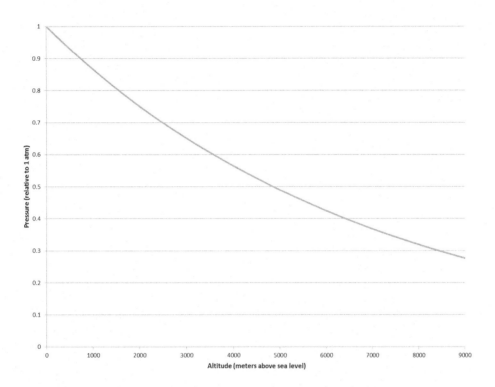

Figure B3. Atmospheric pressure relative to one atmosphere is plotted against height in meters above sea level.

We note from the graph that the pressure at a height 5000 meters above sea level is slightly less than .5 atm (relative to 1 atm).

This should not surprise us since P_{atm} when h = 5 km can be evaluated:

$P_{atm} = 1 * \exp(-5/7) = .48954$ atmosphere.

5. Finding the impact center from the perpendicular bisector of chords on a circle:

Some might recall a theorem from geometry that tells us that the perpendicular bisector of a circle's chord passes through its center. Let us apply that notion to the impact crescent, shown below.

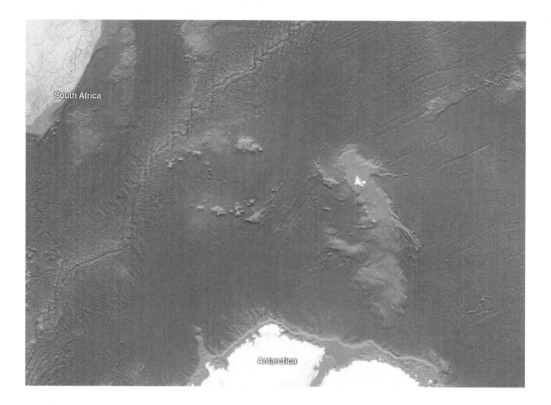

Figure B4. The impact location is shown in this map.

Onto this map we draw two chords along the circular arc, one on each part of the noticeable crescent.

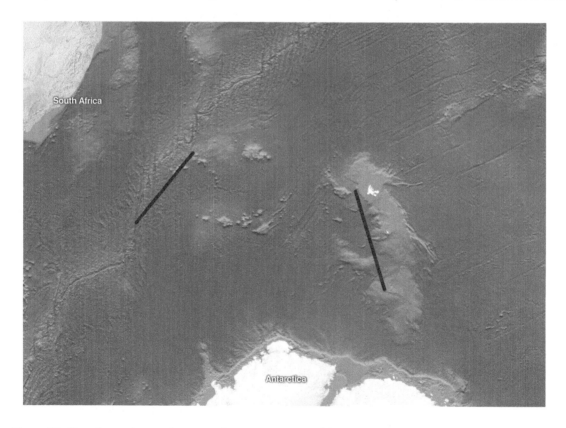

Figure B5. Chords are drawn along circular arc segments of the impact crescent.

Now we draw bisectors perpendicular to these chords.

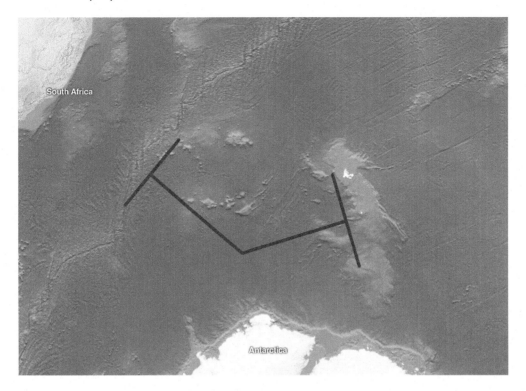

Figure B6. The perpendicular bisectors of the two chords indicates the IO impact center.

The intersection of the two perpendicular bisectors is the circle's center, which in this case is the impact center. Note that the trough carved by the IO's solid nucleus emanates from this central region.

References

A'Hearn, M.F., M.J.S. Belton, W.A. Delamere, J. Kissel, K.P. Klaasen, L.A. McFadden, K.J. Meech, H.J. Melosh, P.H. Schultz, J.M. Sunshine, P.C. Thomas, J. Veverka, D.K. Yeomans, M.W. Baca, I. Busko, C.J. Crockett, S.M. Collins, M. Desnoyer, C.A. Eberhardy, C.M. Ernst, T.L. Farnham, L. Feaga, O. Groussin, D. Hampton, S.I. Ipatov, J.-Y. Li, D. Lindler, C.M. Lisse, N. Mastrodemos, W.M. Owen Jr., J.E. Richardson, D.D. Wellnitz, and R.L. White. 2005. Deep Impact: Excavating comet Tempel 1. *Science* (310) 5746: 258–264.

Allan, D.S. and J.B. Delair. 1997. *Cataclysm! Compelling Evidence of a Cosmic Catastrophe in 9500 B.C.* Rochester, Vermont: Bear and Company. Originally published as *When the Earth Nearly Died* (Bath, England: Gateway Books, 1995).

Churchward, J. 2007. *The Lost Continent of Mu.* Kempton, Illinois: Adventures Unlimited Press.

Fildani, A. and W.R. Normark. 2004. Late Quaternary evolution of channel and lobe complexes of Monterey Fan. *Marine Geology* 206 (1–4): 199–223.

Firestone, R.B., A. West, J.P. Kennett, L. Becker, T.E. Bunch, Z.S. Revay, P.H. Schultz, T. Belgya, D.J. Kennett, J.M. Erlandson, O.J. Dickenson, A.C. Goodyear, R.S. Harris, G.A. Howard, J.B. Kloosterman, P.Lechler, P.A. Mayewski, J. Montgomery, R. Poreda, T. Darrah, S.S. Que Hee, A.R. Smith, A. Stich, W.Topping, J.H. Wittke, W.S. Wolbach. 2007. Evidence for an extraterrestrial impact 12,900 years ago that contributed to the megafaunal extinctions of the Younger Dryas cooling. *Proceedings of the National Academy of Sciences* 104:16016-16021.

Firestone, R.B. 2009. The case for the Younger Dryas extraterrestrial impact event: mammoth, megafauna, and Clovis extinction, 12,900 years ago. *Journal of Cosmology* 2:256-285.

Garcia-Castellanos, D., F. Estrada, I. Jiménez-Munt, C. Gorini, M. Fernàndez, J. Vergés, and R. De Vicente. 2009. Catastrophic flood of the Mediterranean after the Messinian salinity crisis. *Nature* 462 (7274): 778–781.

Gleiser, M. 2014. *The Island of Knowledge: The Limits of Science and the Search for Meaning.* Basic Books.

Gould, S.J. 1985. *The Flamingo's Smile: Reflections in Natural History.* New York: W.W. Norton & Company.

Kennett J.P., D.J. Kennett, B.J. Culleton, J.E.A. Tortosa, J.L. Bischoff, T.E. Bunch, I.R. Daniel Jr., J.M. Erlandson, D. Ferraro, R.B. Firestone, A.C. Goodyear, I. Israde-Alcántara, J.R. Johnson, J.F. Jordá Pardo, D.R. Kimbel, M.A. LeCompte, N.H. Lopinot, W.C. Mahaney, A.M.T. Moore, C.R. Moore, J.H. Ray, T.W. Stafford Jr., K.B. Tankersley, J.H. Wittke, W.S. Wolbach, and A. West. 2015. Bayesian chronological analyses consistent with synchronous age of 12,835–12,735 Cal B.P. for Younger Dryas boundary on four continents. *Proceedings of the National Academy of Sciences of the United States of America* 112 (32): E4344–E4353.

Kerr, R.A. 2005. Deep Impact finds a flying snowbank of a comet. *Science* (309) 5741: 1667.

Kinzie C.R., S.S. Que Hee, A. Stich, K.A. Tague, C. Mercer, J.J. Razink, D.J. Kennett, P.S. DeCarli, T.E. Bunch, J.H. Wittke, I. Israde-Alcántara, J.L. Bischoff, A.C. Goodyear, K.B. Tankersley, D.R. Kimbel, B.J. Culleton, J.M. Erlandson, T.W. Stafford, J.B. Kloosterman, A.M.T. Moore, R.B. Firestone, J.E. Aura Tortosa, J.F. Jordá Pardo, A. West, J.P. Kennett, and W.S. Wolbach. 2014. Nanodiamond-rich layer across three continents consistent with major cosmic impact at 12,800 Cal BP. *The Journal of Geology* 122 (5): 475–506.

Korhonen J.V., J.D. Fairhead, M. Hamoudi, K. Hemant, V. Lesur, M. Mandea, S. Maus, M. Purucker, D. Ravat, T. Sazonova, and E. Thebault. 2007. Magnetic Anomaly Map of the World, 1st ed., Commission for the Geological Map of the World, Paris, France.

Kuhn, T.S. 1962. *The Structure of Scientific Revolutions*. Chicago: University of Chicago Press.

Llamas, B., L. Fehren-Schmitz, G. Valverde, J. Soubrier, S. Mallick, N. Rohland, S. Nordenfelt, C. Valdiosera, S.M. Richards, A. Rohrlach, M.I. Romero, I.F. Espinoza, E.T. Cagigao, J.W. Jiménez, K. Makowski, I.S. Reyna, J.M. Lory, J.A. Torrez, M.A. Rivera, R.L. Burger, M.C. Ceruti, J. Reinhard, R.S. Wells, G. Politis, C.M. Santoro, V.G. Standen, C. Smith, D. Reich, S.Y. Ho, A. Cooper, and W. Haak. 2016. Ancient mitochondrial DNA provides high-resolution time scale of the peopling of the Americas. *Science Advances* 2 (4): e1501385. doi:10.1126/sciadv.1501385.

Mamajek, E.E., S.A. Barenfeld, V.D. Ivanov, A.Y. Kniazev, P. Vaisanen, Y. Beletsky, and H.M.J. Boffin. 2015. The closest known flyby of a star to the solar system. *The Astrophysical Journal Letters* 800 (1): L17.

Metivier F., E. Lajeunesse, and M. Cacas. 2005. Submarine canyons in the bathtub. *Journal of Sedimentary Research* 75 (1): 6–11.

National Oceanic and Atmospheric Administration, Curators of Marine and Lacustrine Geological Samples Consortium: Index to Marine and Lacustrine Geological Samples (IMLGS). National Centers for Environmental Information, leg S1579NC. doi:10.7289/V5H41PB8, accessed November 2011.

Rackham, H. (transl). 1938. Pliny the Elder *Natural History* (London); vol ii, p 91.

Ross N., T. Jordan, R. Bingham, H. Corr, F. Ferraccioli, A. Le Brocq, D. Rippin, A. Wright, M. Siegert. 2013. The Ellsworth Subglacial Highlands: Inception and retreat of the West Antarctic Ice Sheet. *Bulletin of the Geological Society of America,* 19 Sept 2013.

Sedgwick, A. 1831. Address to the Geological Society of London, on retiring from the President's Chair, February 18.

Settegast, M. 1987. *Plato Prehistorian: 10,000 to 5000 B.C. in Myth and Archaeology.* Cambridge, Massachusetts: The Rotenberg Press.

Som S., R. Buick, J.W. Hagadorn, T.S. Blake, J.M. Perreault, J.P. Harnmeijer, and D.C. Catling. 2016. Earth's air pressure 2.7 billion years ago constrained to less than half of modern levels. *Nature Geoscience*, doi:10.1038/ngeo2713.

Sunshine J.M., O. Groussin, P.H. Schultz, M.F. A'Hearn, L.M. Feaga, T.L. Farnham, and K.P. Klaasen. 2007. The distribution of water ice in the interior of Comet Tempel 1. *Icarus* 190 (2): 284–294.

Sweatman S.B., D. Tsikritsis. 2017. Decoding Göbekli Tepe with archaeoastronomy: What does the fox say? *Mediterranean Archaeology and Archaeometry* 17(1): 233-250.

Wilson, E.K. 2005. An icy dustball in outer space. *Chemical & Engineering News* 83 (37): 12.

Zamora, A. 2017. A model for the geomorphology of the Carolina Bays. *Geomorphology* 282:209-216.

CPSIA information can be obtained
at www.ICGtesting.com
Printed in the USA
BVOW05*2306300617

488089BV00002B/2/P